Studies in Classification, Data Analysis, and Knowledge Organization

Studies in Classification, Data Analysis, and Knowledge Organization is a book series which offers constant and up-to-date information on the most recent developments and methods in the fields of statistical data analysis, exploratory statistics, classification and clustering, handling of information and ordering of knowledge. It covers a broad scope of theoretical, methodological as well as application-oriented articles, surveys and discussions from an international authorship and includes fields like computational statistics, pattern recognition, biological taxonomy, DNA and genome analysis, marketing, finance and other areas in economics, databases and the internet. A major purpose is to show the intimate interplay between various, seemingly unrelated domains and to foster the cooperation between mathematicians, statisticians, computer scientists and practitioners by offering well-based and innovative solutions to urgent problems of practice.

Leonardo Grilli · Monia Lupparelli ·
Carla Rampichini · Emilia Rocco · Maurizio Vichi
Editors

Statistical Models and Methods for Data Science

Editors
Leonardo Grilli
Department of Statistics, Computer Science, Applications "G. Parenti"
University of Florence
Florence, Italy

Monia Lupparelli
Department of Statistics, Computer Science, Applications "G. Parenti"
University of Florence
Florence, Italy

Carla Rampichini
Department of Statistics, Computer Science, Applications "G. Parenti"
University of Florence
Florence, Italy

Emilia Rocco
Department of Statistics, Computer Science, Applications "G. Parenti"
University of Florence
Florence, Italy

Maurizio Vichi
Department of Statistical Sciences
Sapienza University of Rome
Rome, Italy

ISSN 1431-8814 ISSN 2198-3321 (electronic)
Studies in Classification, Data Analysis, and Knowledge Organization
ISBN 978-3-031-30163-6 ISBN 978-3-031-30164-3 (eBook)
https://doi.org/10.1007/978-3-031-30164-3

This Springer imprint is published by the registered company Springer Nature Switzerland AG
The registered company address is: Gewerbestrasse 11, 6330 Cham, Switzerland

Preface

This book offers a collection of papers focusing on methods and models in classification and data analysis. Several research topics are covered, ranging from statistical inference and modeling to clustering and factorial methods, from directional data analysis to time series analysis and small area estimation. Applications deal with new investigations in a variety of relevant fields: medicine, finance, engineering, marketing, and cyber risk, to cite a few.

These contributions are a selection of post-conference papers presented at the 13th meeting of the CLAssification and Data Analysis Group (CLADAG) of the Italian Statistical Society (SIS), organized by the Department of Statistics, Computer Science, Applications "G. Parenti" of the University of Florence (Italy) on September 9–11, 2021. The submitted papers followed a careful review process involving two reviewers per paper. In the end, 14 papers were selected for publication in this volume.

Due to the persistent uncertainty about the COVID-19 epidemic, CLADAG 2021 was entirely online. Despite this unfortunate situation, the Conference was highly participated and vital, as the wide range of contributions here collected shows.

CLADAG, a member of the International Federation of Classification Societies (IFCS), organizes an international scientific meeting every two years devoted to presenting theoretical and applied papers in classification and, more generally, data analysis. The meeting includes advanced methodological research in multivariate statistics, mathematical and statistical investigations, survey papers on the state of the art, real case studies, papers on numerical and algorithmic aspects, and applications in special fields of interest at the interface between classification and data science. The Conference aims to encourage the interchange of ideas in the mentioned research fields and disseminate new findings. CLADAG conferences, initiated in 1997 in Pescara (Italy), were soon considered an attractive information exchange market and became an important meeting point for people interested in classification and data analysis. Traditionally, a selection of the presented papers, fully peer-reviewed, is published in a volume of post-conference proceedings.

The Scientific Committee of the 2021 edition has planned Plenary and Invited Sessions to provide a fresh perspective on the state of the art of knowledge and research in the field. The scientific program of CLADAG 2021 is especially rich.

All in all, it comprises five Keynote Lectures, 26 Invited Sessions promoted by the members of the Scientific Program Committee, 10 Contributed Sessions, and a Plenary Session on Statistical Issues in the COVID-19 pandemic.

We thank all the session organizers for inviting renowned speakers from many countries. We are greatly indebted to the referees for the time spent carefully reviewing the papers collected in this book. Special thanks are due to the members of the Local Organizing Committee and all the people who collaborated with CLADAG 2021. Last but not least, we thank all the authors and participants, without whom the Conference would not have been possible. Above all, we thank all participants who chose this book to share their research findings. We hope this book will contribute to fostering new knowledge in the field.

Florence, Italy
November 2022

Leonardo Grilli
Monia Lupparelli
Carla Rampichini
Emilia Rocco
Maurizio Vichi

Contents

Clustering Financial Time Series by Dependency

Andrés M. Alonso, Carolina Gamboa, and Daniel Peña

Abstract In this paper, we propose a procedure for clustering financial time series by dependency on their volatilities. Our procedure is based on the generalized cross correlation between the estimated volatilities of a time series. Monte Carlo experiments are carried out to analyze the improvements obtained by clustering using the squared residuals instead of the levels of the series. Our procedure was able to recover the original clustering structures in all cases in our Monte Carlo study. Finally, the methodology is applied to a set of financial data.

Keywords Estimated volatilities · Generalized cross correlation · Correlation matrix · Unsupervised classification

1 Introduction

The advancement of information technology has made available a large amount of temporal high-frequency data in a variety of fields, like economy, biology, medicine, meteorology, and demography, among others. This fact has promoted research in cluster analysis to find common structures in a data set. The data are grouped searching for homogeneous groups by maximizing some measure of similarity. A variety of methods have been proposed in the literature to cluster sets of time series (see, for example, Caiado et al. (2006), Caiado et al. (2015), Galeano and Peña (2000), Jeong et al. (2011), Piccolo (1990), Díaz and Vilar (2010), and Lafuente-

A. M. Alonso (✉) · D. Peña
Department of Statistics and Institute Flores de Lemus, Universidad Carlos III de Madrid, 28903 Getafe, Spain
e-mail: andres.alonso@uc3m.es

D. Peña
e-mail: dpena@est-econ.uc3m.es

C. Gamboa
Department of Statistics, Universidad Carlos III de Madrid, 28903 Getafe, Spain
e-mail: 100312917@alumnos.uc3m.es

L. Grilli et al. (eds.), *Statistical Models and Methods for Data Science*, Studies in Classification, Data Analysis, and Knowledge Organization,
https://doi.org/10.1007/978-3-031-30164-3_1

Rego and Vilar (2016)). In those methods, clustering is approached from two different perspectives: in the first one, working directly on the original time series and defining an appropriate metric between them; in the second one, projecting the time series on a small space of characteristics or parameters. An often used method was proposed by Piccolo (1990) in which the distances between the parameters of an autoregressive approximation to each time series are used to form the clusters. The previous methodologies are useful when the time series are independent. However, in many applications this assumption is not met, and ignoring this fact can lead us to group highly dependent time series into different groups or to group independent time series into the same group.

In financial time series, the returns do not present a strong structure in the levels, $cor(X_t, X_s) \approx 0$, but they show higher order dependence, as, for instance, in the squares, $cor(X_t^2, X_s^2) \neq 0$. Thus, returns are often uncorrelated but not independent (see, for example, Tsay (2010)). These characteristics have opened a line of research in the grouping of time series taking into account the similarity of the evolution of the univariate conditional variances. For instance, Otranto (2008) and D'Urso et al. (2013) extended the methods proposed in Piccolo (1990) to Generalized AutoRegressive Conditional Heteroscedasticity (GARCH) models. They use the representation of square disturbances in their AR(∞) form in order to compute a distance measure defined as a function of the autoregressive parameters as in Piccolo (1990).

A few articles have proposed methods for clustering by dependency. Zhang and An (2018) proposed a distance measure based on copulas to measure general dependencies of the time series. Alonso and Peña (2019) introduced the generalized cross-correlation (GCC) metric, which is based on the determinant of a set of cross correlations between two time series until a certain lag k. These two methods assumed that the dependency among the time series is on the levels, and does not consider the case in which the dependency could be shown on the conditional variances. To the best of our knowledge, the only proposal to deal with dependency between volatilities is due to La Rocca and Vitale (2021), who proposed to use the auto-distance correlation function (see Zhou, 2012) to handle both linear and nonlinear dependence structures.

In this work, we present a procedure to cluster time series for linear dependency on the volatilities. We believe that this is the most important case in practice, and we extend the approach presented by Alonso and Peña (2019) in the search for dependencies between the squares of two time series or between the estimated volatilities. The work is organized as follows. Section 2 introduces the notation and reviews some conditional heteroscedastic models. Section 3 extends the GCC measure for measuring dependencies between squares of time series. In Sect. 4, we present some Monte Carlo simulations in order to evaluate the performance of GCC measure to detect relations between squares of heteroscedastic time series. Section 5 illustrates the use of the proposed procedure to a well-known set of 100 financial time series.

2 Conditional Heteroscedastic Models

Volatility, or conditional variance, is an important feature of financial markets, since it captures the conditional variations in the returns of the assets. As discussed in Otranto (2008), it is generally considered to be a proxy of financial risk, since a high volatility means that the value of returns can change dramatically in short periods of time, and low volatility is associated with stable assets or periods. Some other characteristics that usually appear in asset returns are as follows: (i) their distributions are heavy-tailed; (ii) there exist volatility clusters, i.e. periods of high volatility are often followed by periods of low volatility and vice versa; (iii) although the returns are not correlated, their square values have a strong autocorrelation structure (see, for example, Tsay (2010)). Based on these properties, some models have been proposed in the literature for studying the behavior of these time series. Engle (1982) proposed the well known *Autoregressive Conditional Heteroscedasticity* (ARCH) model in which the dependence between squared asset returns is parameterized by using an autoregressive type model and the volatility, σ_t^2, is a function of past squared returns. As often this model requires a high order AR, Bollerslev (1986) proposed the GARCH model where the volatility also depends on past values of the conditional variance. Given a stationary time series $x_t = \mu + e_t$ with mean μ and heteroscedastic noise e_t, it is said that the series follows a GARCH(p,q) model if the noise follows the equation

$$e_t = \sigma_t \epsilon_t,$$

where ϵ_t is an i.i.d. random variable with mean zero and variance one, and σ_t is the volatility that can be expressed as

$$\sigma_t^2 = \omega + \sum_{j=1}^{p} \alpha_j e_{t-j}^2 + \sum_{l=1}^{q} \beta_l \sigma_{t-l}^2,$$

where $\omega > 0$ and $0 \leq \alpha_j, \beta_l < 1$, in order to ensure that the conditional variance is positive, and $\left(\sum_{j=1}^{p} \alpha_j + \sum_{j=1}^{p} \beta_l\right) < 1$ since the process is stationary.

Those models have been generalized to the multivariate framework but, as mentioned by Tsay (2014), this approach suffers the curse of dimensionality. For instance, with n time series we have n conditional variances and $n(n-1)/2$ conditional covariances. Our work would make it possible to reduce the number of elements to be modeled, as the covariances between groups of independent (or very slightly dependent) series can be considered negligible.

3 Procedure for Clustering Time Series by Dependency

Alonso and Peña (2019) proposed a measure based on the dependency between the conditional means of two time series. Our goal is to extend this measure to the dependency between conditional variances, σ_t^2. The procedure could be extended in order to find interesting relations between other conditional moments of time series.

Let w_t and z_t be two stationary time series, and let $x_t = w_t^2$, $y_t = z_t^2$ be their corresponding squares that will also be stationary. Using the results given in Alonso and Peña (2019), we are going to define a dependence measure between (x_t, y_t). Suppose that we have standardized the squared series so that $E(x_t) = E(y_t) = 0$ and $var(x_t) = var(y_t) = 1$. For lags $h = 0, \pm 1, \cdots, \pm k$, the autocorrelation function of x_t is $\rho_x(h) = E(x_{t-h}x_t)$ and the cross correlation between x_t and y_t is $\rho_{xy}(h) = E(x_{t-h}y_t)$. The linear dependency between the two time series of squares can be summarized in the matrix

$$\mathbf{R}(h) = \begin{pmatrix} \rho_x(h) & \rho_{xy}(h) \\ \rho_{yx}(h) & \rho_y(h) \end{pmatrix}.$$

For the set of lags from 0 to k, we define the matrix,

$$\mathbf{R}_k = \begin{pmatrix} \mathbf{R}(0) & \mathbf{R}(1) & \dots & \mathbf{R}(k) \\ \mathbf{R}(-1) & \mathbf{R}(0) & \dots & \mathbf{R}(k-1) \\ \vdots & \vdots & \ddots & \vdots \\ \mathbf{R}(-k) & \mathbf{R}(-k+1) & \dots & \mathbf{R}(0) \end{pmatrix},$$

which corresponds to the correlation matrix of the stationary process

$$(x_t, y_t, x_{t-1}, y_{t-1}, \dots, x_{t-k}, y_{t-k})'.$$

We can reorganize the elements of the above vector as $Z_{t,2(k+1)} = (X'_{t,(k+1)}, Y'_{t,(k+1)})'$, where $X'_{t,(k+1)} = (x_t, \cdots, x_{t-k})'$ and $Y'_{t,(k+1)} = (y_t, \cdots, y_{t-k})'$. In this case, the correlation matrix will be

$$\mathbf{R}_{yx,k} = \begin{pmatrix} \mathbf{R}_{yy,k} & \mathbf{C}_{xy,k}^T \\ \mathbf{C}_{xy,k} & \mathbf{R}_{xx,k} \end{pmatrix},$$

where $\mathbf{R}_{xx,k}$ and $\mathbf{R}_{yy,k}$ are the correlation matrices for the $X_{t,k}$ and $Y_{t,k}$ processes, respectively, and $\mathbf{C}_{xy,k}$ the matrix of cross correlations between these two vectors.

$\mathbf{R}_{yx,k}$ matrix verifies that $0 \leq \det(\mathbf{R}_{yx,k}) \leq 1$. The determinant will take the value one when the matrix is diagonal, i.e. when the two time series do not have cross correlations or autocorrelations different than zero for $k \neq 0$; and the determinant will take the value zero when there is a perfect linear dependence between the components of the vector $Z_{t,2(k+1)}$.

In Alonso and Peña (2019) is shown that the $\det(\mathbf{R}_{yx,k})$ by itself is not sufficient to capture the linear relationship between the two time series, since it depends on the cross correlations as well as the autocorrelation of both processes. For example, if $C_{xy,k} = 0$ then $\det(R_{xy,k}) = \det(R_{xx,k})\det(R_{yy,k})$; in addition, if one of the series presents strong autocorrelations we will have $\det(R_{yx,k})$ to be close to zero, regardless of whether the series are independent or not. For this reason, an alternative similarity measure was proposed, described as follows:

$$GCC(x_t, y_t) = 1 - \left(\frac{\det(\mathbf{R}_{yx,k})}{\det\left(\mathbf{R}_{xx,k}\right)\det\left(\mathbf{R}_{yy,k}\right)} \right)^{1/(k+1)}.$$

This similarity measure $GCC(x_t, y_t)$ satisfies the following properties: (1) $GCC(x_t, y_t) = GCC(y_t, x_t)$; (2) $0 \leq GCC(y_t, x_t) \leq 1$, it takes the one value in the case that the dependence between both variables is perfectly linear, and takes the value zero in the case that all cross-correlation coefficients are zero. The last statement was proven at Alonso and Peña (2019, Eq. 11) using Hadamard's inequality. It is called *Generalized Cross-Correlation* measure (GCC). In addition, this measure takes into account the dimension of the matrices and thus allows us to compare matrices of different dimensions. The selection of k can be done using the BIC criteria (see Sect. 5.1 in Alonso and Peña 2019). Additionally, the analyst can have a priori information about which lags are relevant to her specific problem.

Based on this measure, a dissimilarity between x_t and y_t can be defined by $d(x_t, y_t) = 1 - GCC(x_t, y_t)$; in that way, high dissimilarity values are associated with weak linearity dependencies, and values close to zero will be related to strong linear dependencies between x_y and y_t. Once the pairwise dissimilarity between time series is obtained, we can apply any clustering method that uses dissimilarity matrices as input. In this work, we will use a hierarchical clustering algorithm with single linkage as this allows us to find interesting dependency structures such as chain dependency (see details in Alonso and Peña (2019) and Alonso et al. (2021)).

4 Simulation Study

In this section, we carry out a Monte Carlo experiment to evaluate the performance of the proposed methods using ARCH(1) models. Suppose that

$$\begin{cases} \epsilon_{t,x} = S_{t,x}\eta_{t,x} \\ \epsilon_{t,y} = S_{t,y}\eta_{t,y} \end{cases}, \tag{1}$$

where $S_{t,y}$, $S_{t,x}$, $\eta_{t,y}$, $\eta_{t,x}$ are pairwise independent, except by the pair $(\eta_{t,y}, \eta_{t,x})$ that will be dependent. The dummy variables $S_{t,y}$ and $S_{t,x}$ take only two possible values $\{-1, 1\}$, each with probability 1/2 and are independent over time, and $\eta_{t,y}$

and $\eta_{t,x}$ are standardized white noise normal variables with $\text{cor}(\eta_{t,y}, \eta_{t,x}) = \rho > 0$. Then, the variables $\epsilon_{t,x}$ and $\epsilon_{t,y}$ have mean zero, variance one, and $\text{cor}(\epsilon_{t,x}^2, \epsilon_{t,y}^2) = \text{cor}(\eta_{t,x}^2, \eta_{t,y}^2) = \rho^2$ (see Appendix). Furthermore, we have

$$\begin{cases} E(\epsilon_{t,x}) = E(S_{t,x}\eta_{t,x}) = E(S_{t,x})E(\eta_{t,x}) = 0 \\ E(\epsilon_{t,y}) = E(S_{t,y}\eta_{t,y}) = E(S_{t,y})E(\eta_{t,y}) = 0 \end{cases} . \tag{2}$$

Thus, taking into account (2), we obtain

$$\text{cov}(\epsilon_{t,x}, \epsilon_{t,y}) = E(\epsilon_{t,x}\epsilon_{t,y}) = E(S_{t,x}S_{t,y}\eta_{t,x}\eta_{t,y}) = 0.$$

Thus, the series will have linear dependencies on their squares, but will be uncorrelated on the levels.

Consider a data set $e_{t,i} = \sigma_{t,i}\epsilon_{t,i}$, with $i = 1, ..., 15$ and each of the 15 time series having size $T = 1000$. The $\sigma_{t,i}$ follow the ARCH(1) model, $\sigma_{t,i}^2 = \alpha_{0,i} + \alpha_{1,i}e_{t-1}^2$ and

$$\begin{cases} \text{Model 1:}\alpha_{0,i} = 0.01 \text{ and } \alpha_{1,i} = 0.3 \quad \text{for } i \in \{1, \cdots, 5\} \cup \{11, \cdots, 15\} \\ \text{Model 2:}\alpha_{0,i} = 0.001 \text{ and } \alpha_{1,i} = 0.6 \text{ for } i \in \{6, \cdots, 10\} \end{cases}$$

Using (1), we introduce the cross dependence in the set of time series $\{e_{t,i}\}$ through the noise $\{\epsilon_{t,i}\}$. For this, we will assume that the $\eta_{t,i}$ variables are multivariate normal with dependence structure defined by $\rho(i, j) = \text{cor}(\eta_{t,i}, \eta_{t,j})$. We will consider five scenarios where there are two (or six) clusters by dependency. Different types of dependency will be studied: independence, chain dependency, and full dependency.

The noise in the five scenarios is defined as follows:

- Scenario 1: $\rho(i, j) = 0.9$ for $i = 1, \ldots, 9$ and $j = i + 1, \ldots, 10$, and 0 otherwise. In this scenario, we have ten full dependent series and five independent ones; in total we have six groups.
- Scenario 2: $\rho(i, j) = 0.9$ for $i = 1, \ldots, 9$ and $j = i + 1, \ldots, 10$, and $\rho(i, i + 1) = 0.5$ for $i = 11, \ldots, 14$, and 0 otherwise. In this scenario, we have ten full dependent series and five chain-dependent ones; in total we have two groups.
- Scenario 3: $\rho(i, j) = 0.9$ for $i = 1, \ldots, 9$ and $j = i + 1, \ldots, 10$. $\rho(i, j) = 0.9$ for $i = 11, \ldots, 14$ and $j = i + 1, \ldots, 15$, and 0 otherwise. In this scenario, we have ten full dependent series and five full dependent ones; in total we have two groups.
- Scenario 4: $\rho(i, i + 1) = 0.5$ for $i = 1, \ldots, 9$ and $\rho(i, i + 1) = 0.5$ for $i = 11, \ldots, 14$, and 0 otherwise. In this scenario, we have ten chain- dependent series and five chain-dependent ones; in total we have two groups.
- Scenario 5: $\rho(i, i + 1) = 0.5$ for $i = 1, \ldots, 9$, and 0 otherwise. In this scenario, we have ten chain-dependent series and five independent ones; in total we have six groups.

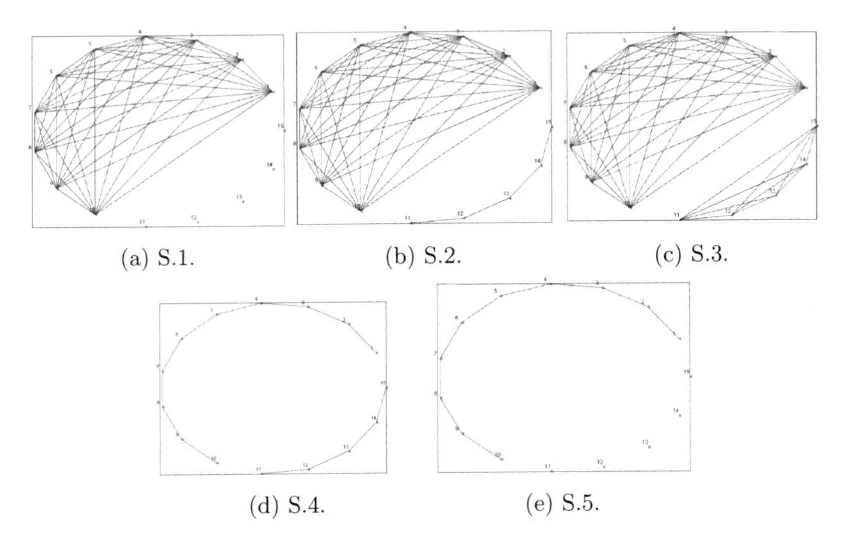

(a) S.1. (b) S.2. (c) S.3.

(d) S.4. (e) S.5.

Fig. 1 Dependence structure representation

In Fig. 1, we present these five scenarios; each one is represented by an ellipse where each point corresponds to a time series. The line represents the not null cross correlation between two time series.

For the clustering procedure, we are going to compare the results using the levels and the squares of the time series. The first, the bivariate $GCC(e_{i,t}, e_{j,t})$ measure on the levels, will be denoted by GCC and the second, the GCC measure on the squares of $e_{i,t}$ that is $GCC(e_{i,t}^2, e_{j,t}^2)$, will be denoted by GCC2. We have also computed the GCC measure on the estimated volatility that is $GCC(\hat{\sigma}_{t,x}^2, \hat{\sigma}_{t,x}^2)$ (denoted by GCC_vol), where the volatilities have been estimated by using a GARCH(1,1) which is a usual selection in financial time series modeling. Note that the results of using the estimated volatility will depend on the chosen model, and this is a strong limitation. Using the squares is equivalent to estimating the volatility using an ARCH(1) process, as then the volatilities are linear transformations of the squared residuals.

As proposed by Hubert and Arabie (1985), the partitions can be compared using *Adjusted Rand Index (ARI)*. ARI is based on counting pairs of observations that are classified in the same cluster under two cluster partitions. The closer the index is to one (zero), the higher (lower) the agreement between the two partitions. Table 1 reports the means of the ARI measure from 100 replicates for these five scenarios. We group time series using hierarchical clustering with a single linkage. Assuming that the number of groups is known, we can observe that when the estimated volatility is used, the performance of this methodology is similar to the GCC2. Both procedures capture the grouping structure in all cases for scenarios 1 to 3. Also, both measures have a better performance than GCC measure that uses the original time series. Additional Monte Carlo experiments have been performed, but are not presented here due to space constraints.

Table 1 Clustering performance ARCH(1) model

Method	Scenario 1	Scenario 2	Scenario 3	Scenario 4	Scenario 5
GCC	0.70	0.16	0.24	0.01	0.09
GCC2	1.00	1.00	1.00	0.89	0.90
GCC_vol	1.00	0.99	1.00	0.89	0.90

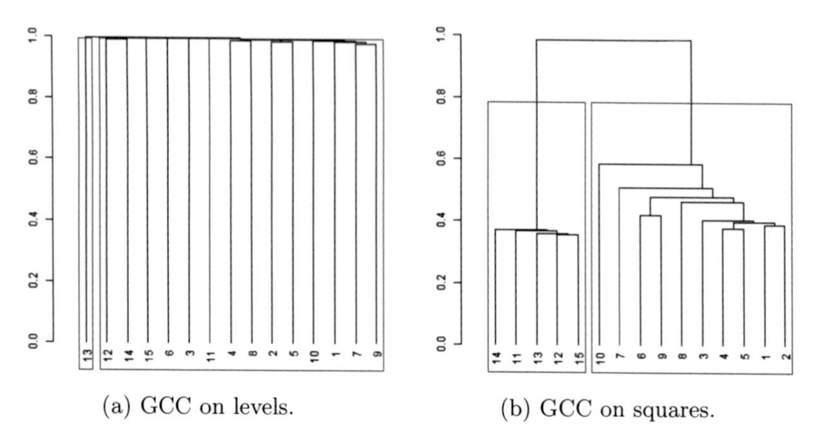

(a) GCC on levels. (b) GCC on squares.

Fig. 2 Example of dendrograms (using single linkage) for scenario S.3

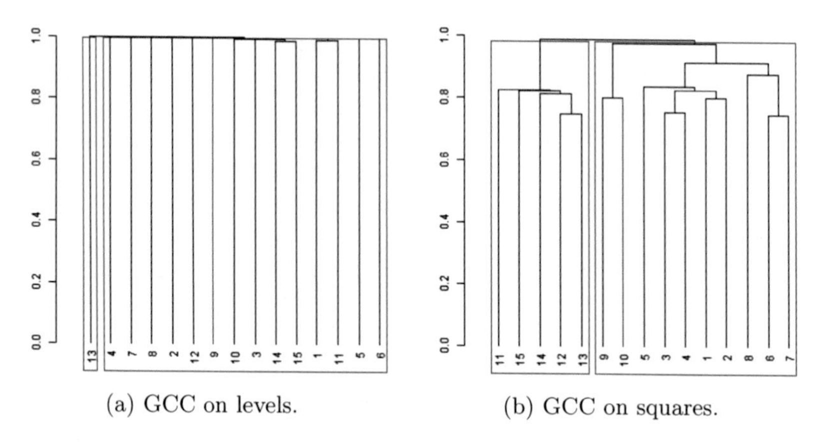

(a) GCC on levels. (b) GCC on squares.

Fig. 3 Example of dendrograms (using single linkage) for scenario S.4

Figures 2 and 3 show an example of dendrograms obtained using the GCC measure for the third and fourth scenarios when we use the original time series (levels) and their squares. The two groups for scenario 3 are clearly distinguishable in the dendrograms built from the squares; besides, we note that this structure cannot be captured when we use levels. For the other scenarios, we obtain similar results, however, when there is chain dependence, the unions of the groups are presented at high points in the dendrogram (see Fig. 3).

5 Real Data Example

In this section, we will analyze 100 time series that contain equal-weighted returns from the intersections of five portfolios formed on size (market equity, ME) corresponding to Europe (UE), Japan (JAP), Pacific Asia (PA)—except Japan, and North America (AM); and five portfolios formed on the ratio of book equity to market equity (BE/ME). The size is coded from 1(small) to 5(big) and the book to market ratio is coded from 1(low) to 5(high). A detailed description can be found on the web page developed by Kenneth R. French http://mba.tuck.dartmouth.edu/pages/faculty/ken.french/data_library.html. The data are taken from July 1990 to April 2019 making a total of 7522 days.

In Fig. 4, we present the dendrogram obtained by using a hierarchical clustering procedure with a single linkage over the levels of daily returns. The number of clusters have been obtained by using the Silhouette statistic (see Rousseeuw (1987)). This statistic measures the mean similarity (closeness) of observations in its own group to the similarity with other groups. The Silhouette statistic detects four clusters in the data set which correspond to the four regions analyzed. In addition, the series that belong to each cluster present a strong linear dependence between them, except in the cluster of the Asian time series. When we use the squares of daily returns for clustering, the Silhouette statistic detects five clusters. They are the four main groups found in levels and a fifth group composed by a single time series belonging to Pacific Asia. Also, in the squares the group of American, time series presents weaker dependencies than in levels (see Fig. 5).

In Fig. 6, we show that the dependency structures based on levels differ from the one based on squared returns. In particular, it is remarkable that portfolios AM12, AM13, AM14, AM15 make up a group of dependent series on the levels, however, this group is divided when squares are taken into account, the same is true for the group of portfolios AM52, AM53, AM54.

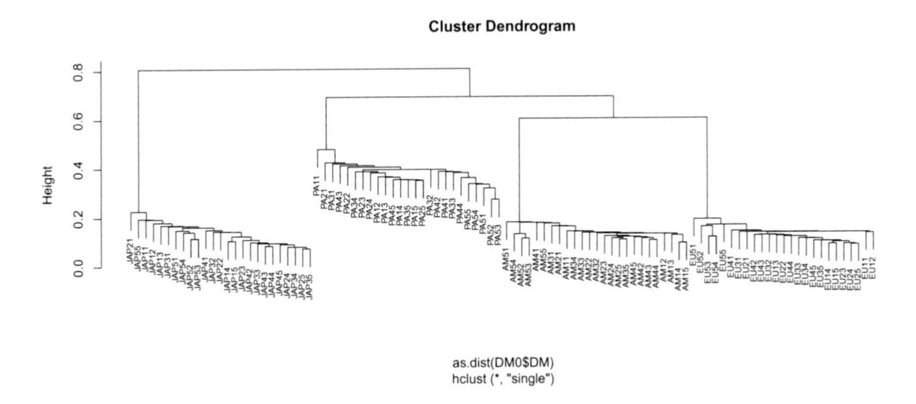

Fig. 4 Dendrogram using GCC on returns from 100 portfolios (using single linkage method)

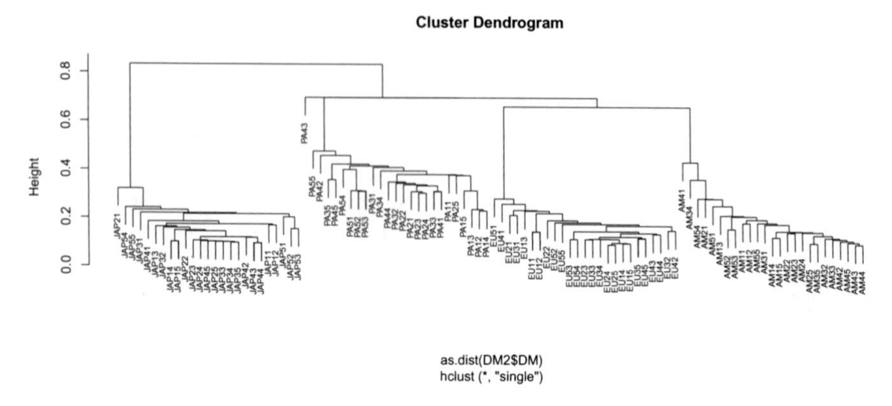

Fig. 5 Dendrogram using GCC on squares of returns from 100 portfolios (using single linkage method)

The results with the estimated volatilities are similar to the ones obtained with the square returns, so we omit the figure but it is available upon request to the authors.

6 Conclusions

Clustering time series in high-dimensional dynamic data sets by taking into account their dependency is an important field of research. In financial time series, the dependency appears often in the volatilities, and clustering the series for similar volatility can be very helpful to build factorial models for volatilities in large data sets. The use of the generalized cross-correlation measure applied to the dependencies on the squares of the series, or on their estimated volatilities, has been shown to be very useful in the Monte Carlo study. Also, some interesting features in the real data analysis have been illustrated by comparing the clusters in levels and squares. We conclude that clustering using the linear dependency between the volatilities seems to be a promising line of research although other possible ways of nonlinear dependency in real data should be explored in the future.

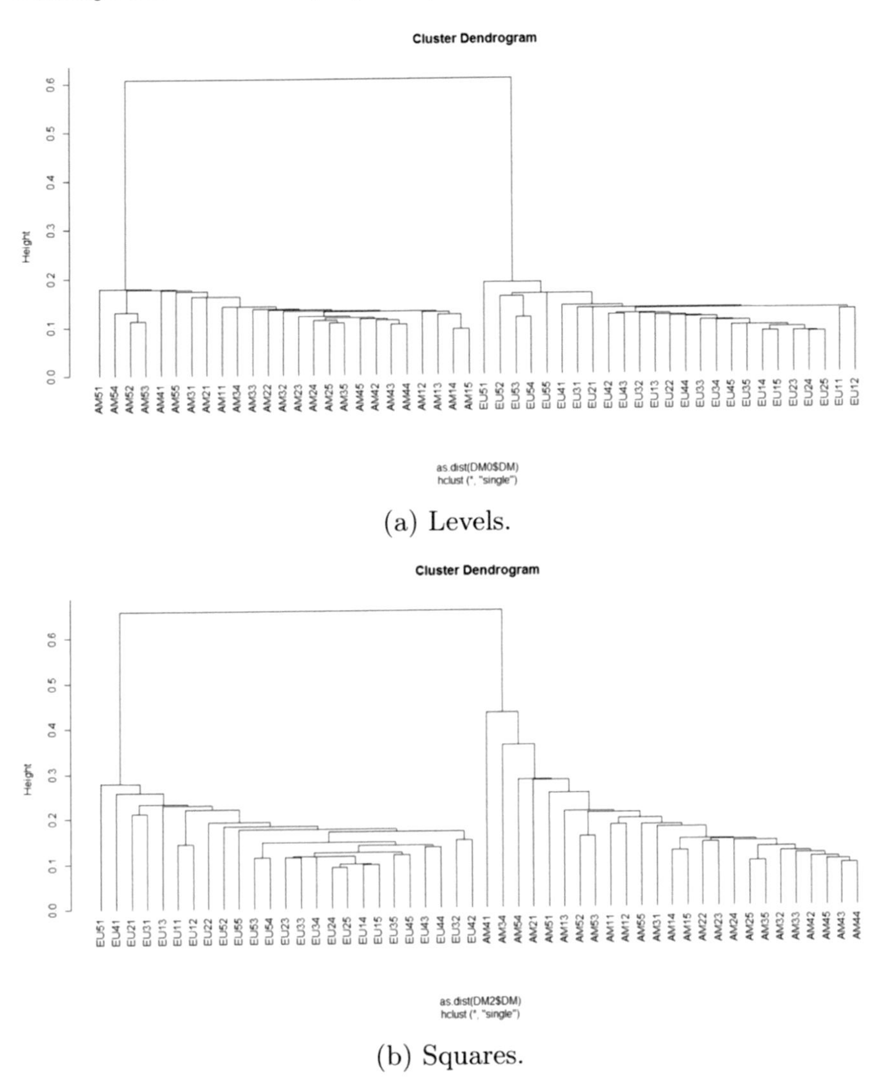

(a) Levels.

(b) Squares.

Fig. 6 Dendrogram using GCC of returns and squares from American and European portfolios

Acknowledgements The authors gratefully acknowledge the financial support from the Spanish government Agencia Estatal de Investigación (PID2019-108311GB-I00/AEI/10.13039/501100011033) and Comunidad de Madrid.

Appendix

Correlation of Two Dependent Chi-square Variables

Let $\eta_{t,x}$ and $\eta_{t,y}$ be normal variables with mean 0, variance 1, and $\text{cor}(\eta_{t,x}, \eta_{t,y}) = \rho$. Suppose that $\eta_{t,y} = \rho\eta_{t,x} + a\eta_{t,z}$ with $\eta_{t,z}$ a standard normal variable independent of $\eta_{t,x}$ and $a^2 = 1 - \rho^2$. Thus

$$\eta_{t,x}^2\eta_{t,y}^2 = \eta_{t,x}^2(\rho^2\eta_{t,x}^2 + 2\rho a\eta_{t,x}\eta_{t,z} + a^2\eta_{t,z}^2) = \eta_{t,x}^4\rho^2 + 2a\rho\eta_{t,x}^3\eta_{t,z} + a^2\eta_{t,z}^2\eta_{t,x}^2.$$

Using the moments of standard normal variables and the independence between $\eta_{t,x}$ and $\eta_{t,z}$, we have that $E(\eta_{t,x}^2\eta_{t,y}^2) = 3\rho^2 + a^2 = 2\rho^2 + 1$. Thus, we conclude that

$$\text{cov}(\eta_{t,x}^2, \eta_{t,y}^2) = E(\eta_{t,x}^2\eta_{t,y}^2) - E(\eta_{t,x}^2)E(\eta_{t,y}^2) = E(\eta_{t,x}^2\eta_{t,y}^2) - 1 = 2\rho^2.$$

On the other hand, we know that $var(\eta_{t,x}^2) = var(\eta_{t,y}^2) = 2$, therefore,

$$\text{cor}(\eta_{t,x}^2, \eta_{t,y}^2) = \frac{2\rho^2}{2} = \rho^2.$$

References

Alonso, A. M., & Peña, D. (2019). Clustering time series by linear dependency. *Statistics and Computing, 29*, 655–676. https://doi.org/10.1007/s11222-018-9830-6.

Alonso, A. M., D'Urso, P., Gamboa, C., & Guerrero, V. (2021). Cophenetic-based fuzzy clustering of time series by linear dependency. *International Journal of Approximate Reasoning, 137*, 114–136. https://doi.org/10.1016/j.ijar.2021.07.006.

Bollerslev, T. (1986). Generalized autoregressive conditional heteroskedasticity. *Journal of Econometrics, 31*, 307–327. https://doi.org/10.1016/0304-4076(86)90063-1.

Caiado, J., Crato, N., & Peña, D. (2006). A periodogram-based metric for time series classification. *Computational Statistics & Data Analysis, 50*, 2668–2684. https://doi.org/10.1016/j.csda.2005.04.012.

Caiado, J., Maharaj, E. A., & D'Urso, P. (2015). Time-series clustering. In: *Handbook of cluster analysis* (pp. 262–285). Chapman and Hall/CRC. https://doi.org/10.1201/9780429058264.

Díaz, S. P., & Vilar, J. A. (2010). Comparing several parametric and nonparametric approaches to time series clustering: A simulation study. *Journal of Classification, 27*, 333–362. https://doi.org/10.1007/s00357-010-9064-6.

D'Urso, P., Cappelli, C., Di Lallo, D., & Massari, R. (2013). Clustering of financial time series. *Physica A: Statistical Mechanics and its Applications, 392*, 2114–2129. https://doi.org/10.1016/j.physa.2013.01.027.

Engle, R. F. (1982). Autoregressive conditional heteroscedasticity with estimates of the variance of United Kingdom inflation. *Econometrica, 50*, 987–1007. https://doi.org/10.2307/1912773.

Galeano, P., & Peña, D. (2000). Multivariate analysis in vector time series. *Resenhas do Instituto de Matemática e Estatística da Universidade de S ao Paulo, 4*, 383–403.

Hubert, L., & Arabie, P. (1985). Comparing partitions. *Journal of Classification, 2*, 193–218. https://doi.org/10.1007/BF01908075.

Jeong, Y.-S., Jeong, M. K., & Omitaomu, O. A. (2011). Weighted dynamic time warping for time series classification. *Pattern Recognition, 44*, 2231–2240. https://doi.org/10.1016/j.patcog.2010.09.022.

Lafuente-Rego, B., & Vilar, J. A. (2016). Clustering of time series using quantile autocovariances. *Advances in Data Analysis and Classification, 10*, 391–415. https://doi.org/10.1007/s11634-015-0208-8.

La Rocca, M., & Vitale, V. (2021). Clustering time series by nonlinear dependence. In M. Corazza et al. (Eds.), *Mathematical and Statistical Methods for Actuarial Sciences and Finance* (pp. 291–297). https://doi.org/10.1007/978-3-030-78965-7_43.

Otranto, E. (2008). Clustering heteroskedastic time series by model-based procedures. *Computational Statistics & Data Analysis, 52*, 4685–4698. https://doi.org/10.1016/j.csda.2008.03.020.

Piccolo, D. (1990). A distance measure for classifying ARIMA models. *Journal of Time Series Analysis, 11*, 153–164. https://doi.org/10.1111/j.1467-9892.1990.tb00048.x.

Rousseeuw, P. J. (1987). Silhouettes: A graphical aid to the interpretation and validation of cluster analysis. *Journal of Computational and Applied Mathematics, 20*, 53–65. https://doi.org/10.1016/0377-0427(87)90125-7.

Tsay, R. S. (2010). *Analysis of financial time series*. Wiley. https://doi.org/10.1002/9780470644560.

Tsay, R. S. (2014). *Multivariate time series analysis*. Wiley. https://doi.org/10.1002/9780470644560.ch8.

Zhang, B., & An, B. (2018). Clustering time series based on dependence structure. *PloS One, 13*, e0206753. https://doi.org/10.1371/journal.pone.0206753.

Zhou, Z. (2012). Measuring nonlinear dependence in time-series, a distance correlation approach. *Journal of Time Series Analysis, 33*, 438–457. https://doi.org/10.1111/j.1467-9892.2011.00780.x.

The Homogeneity Index as a Measure of Interrater Agreement for Ratings on a Nominal Scale

Giuseppe Bove

Abstract Interrater agreement for classifications on nominal scales is usually evaluated by overall measures across targets (subjects or objects) like Cohen's kappa index. In this paper, the homogeneity index for a qualitative variable is proposed to evaluate the agreement between raters for each target, and also to obtain a global measure of interrater agreement for the whole group of targets evaluated. The target-specific and global measures proposed do not depend on a particular definition of agreement (simultaneously between two, three, or more raters) and are not influenced by marginal rater distributions of scale like most kappa-type indices.

Keywords Nominal classification scales · Interrater agreement · Homogeneity index

1 Introduction

In behavioral and biomedical sciences, classifications of subjects or objects (targets) into predefined classes or categories and the analysis of their agreement are a rather common activity. For instance, agreement between clinical diagnoses provided by more physicians (raters) who recognize symptoms with respect to the type of disease is considered for identifying the best treatment for patients. The patient will not feel confident if physicians seriously differ in opinion. The rating procedure (or scale) can be used with confidence and will be accepted by the patient depending on the extent to which the diagnoses coincide. This does not only apply to physicians but also in general to all those situations where it is impossible to establish 'truth classification' in an objective way. Hence, in this type of application it is important to analyze interrater absolute agreement, that is, the extent to which raters assign the same category on the rating scale. Absolute agreement is distinct from association.

G. Bove (✉)
Dipartimento di Scienze della Formazione, Università degli Studi Roma Tre, Rome, Italy
e-mail: giuseppe.bove@uniroma3.it

L. Grilli et al. (eds.), *Statistical Models and Methods for Data Science*, Studies in Classification, Data Analysis, and Knowledge Organization,
https://doi.org/10.1007/978-3-031-30164-3_2

Strong absolute agreement requires strong association, but strong association can exist without strong absolute agreement. Consequently, a measure of association is not necessarily a good measure of absolute agreement. In the following, attention is restricted to the case of a nominal scale, and ordinal or interval scales will not be considered. A consequence of this restriction is that agreement statistics based on Pearson product moment correlations or upon analysis of variance techniques (which assume interval categories) are ruled out.

1.1 Aims of This Contribution

Agreement between two raters who rate each of a sample of targets on a nominal scale is usually assessed with Cohen's kappa (Cohen, 1960). Generalizations of kappa for more than two raters and for targets assessed by different groups of raters have been proposed by many authors (e.g., Fleiss, 1971; Conger, 1980). These indices are used to analyze the agreement between multiple raters for a whole group of targets. Moreover, methods to detect subsets of raters who demonstrate a high level of interobserver agreement were considered, for instance, by Landis and Koch (1977). Less frequent agreement on a single target has been considered (O'Connell, & Dobson, 1984), in spite of the fact that having evaluations of agreement on a single case is particularly useful, for example, in situations where the rating scale is being tested, and it is necessary to identify any changes to improve it, or to request from the raters a specific comparison on a single case in which the agreement is poor. Another situation that needs the analysis of agreement on single targets is when teachers in a school are evaluated by a questionnaire administered to different groups of raters (pupils, peers, principal). In this case, comparisons are made between the level of agreement in each group for each single teacher and for the whole group of teachers. Many of the indices available are not able to measure the agreement when each target (teacher) is evaluated by a completely different set of raters (pupils) or by a different number of raters.

In the next sections, after a brief review of the indices proposed to measure agreement for nominal scales, an index to measure the interrater agreement on single targets is proposed based on a measure of dispersion for nominal variables. Furthermore, a global measure of agreement on the whole group of targets obtained as the arithmetic average of the target values of the index will be also considered and applied to a data set concerning the cause of death of 35 hypertensive patients. The measures proposed allow avoiding paradoxes caused by the dependence of most kappa-type indices on marginal distributions of the categories of the scale observed for each rater.

1.2 Measures of Interrater Agreement for Nominal Scales

Ratings provided by a group of raters for a group of targets have been presented in different ways. In the case of two raters, a contingency table with raters considered as two crossed variables, having the levels of the scale as categories, and entries in each cell given by the number (or proportion) of ratings corresponding to each combination of categories is the most common data representation (e.g., Cohen, 1960, Table 1).

For multiple (two or more) raters, the case mainly considered in this paper, data are usually represented either 1) as multivariate matrix targets × raters (see Tables 1 or 2) as a distribution of raters by targets and category (see Table 2), according to the type of measure of interrater agreement to be computed.

In Table 1, x_{ig} is the category that rater g assigns to target i. The approach mainly considered to extend coefficient kappa (Cohen, 1960) to the case of multiple raters is based on the representation in Table 2, where r_{ik} is the number of raters who classified target i into category k, r_i is the number of raters who rated target i (equal to R if

Table 1 Targets × Raters matrix of ratings

Targets	Rater 1	Rater 2	…rater g …	Rater R
1	x_{11}	x_{12}	$\ldots x_{1g} \ldots$	x_{1R}
2	x_{21}	x_{22}	$\ldots x_{2g} \ldots$	x_{2R}
⋮	⋮	⋮	⋮	⋮
i	x_{i1}	x_{i2}	$\ldots x_{ig} \ldots$	x_{iR}
⋮	⋮	⋮	⋮	⋮
N	x_{N1}	x_{N2}	$\ldots x_{Ng} \ldots$	x_{NR}

Table 2 Targets × Categories distribution of raters with column averages and row totals

Targets	Categ. 1	Categ. 2	…categ. k …	Categ. q	Total
1	r_{11}	r_{12}	$\ldots r_{1k} \ldots$	r_{1q}	r_1
2	r_{21}	r_{22}	$\ldots r_{2k} \ldots$	r_{2q}	r_2
⋮	⋮	⋮	⋮	⋮	⋮
i	r_{i1}	r_{i2}	$\ldots r_{ik} \ldots$	r_{iq}	r_i
⋮	⋮	⋮	⋮	⋮	⋮
N	r_{N1}	r_{N2}	$\ldots r_{Nk} \ldots$	r_{Nq}	r_N
Average	$\bar{r}_{.1}$	$\bar{r}_{.2}$	$\ldots \bar{r}_{.k} \ldots$	$\bar{r}_{.q}$	$\bar{r}$

all raters rate target i), $\overline{r}_{.k}$ is the mean number of raters who classified a target into category k, and $\overline{r}$ is the mean number of raters who classified a target (equal to R if all raters rate each target). Extensions of Cohen's kappa (Cohen, 1960) are mostly formulated as

$$kappa = \frac{P_a - P_e}{1 - P_e},$$

where P_a is the observed proportion of agreement and P_e is the proportion of chance agreement. The idea of adjusting P_a for chance agreement P_e in kappa is because in some situations raters could be unclear about the categorization of a target and they could agree by pure chance. The denominator $(1 - P_e)$ is the maximum value assumed by $(P_a - P_e)$. Most extensions share the same definition of P_a, but differ on the expression to compute chance agreement. The extent of agreement among the r_i raters for the ith target is usually considered as the proportion of agreeing pairs out of all the $r_i(r_i - 1)/2$ possible pairs of assignments. This proportion in the case of balanced data ($r_i = R, i = 1, 2, .., N$) is

$$\sum_{k=1}^{q} \frac{r_{ik}(r_{ik} - 1)}{R(R - 1)}.$$

The overall proportion of agreement P_a is the arithmetic mean of the proportions of agreement of the N targets

$$P_a = \frac{1}{N} \sum_{i=1}^{N} \sum_{k=1}^{q} \frac{r_{ik}(r_{ik} - 1)}{R(R - 1)}.$$

For only two raters (the case considered in Cohen (1960)), P_a is directly interpretable as the proportion of joint judgments in which there is agreement (the sum of the proportions positioned along the main diagonal of the contingency table with raters considered as the two crossed variables). We remark here that the agreement measured by P_a is between pairs of raters (2-agreement type), but joint agreement between three (3-agreement type), four (4-agreement type), or more raters could be considered as well (Conger, 1980; Warrens, 2012). The definition of the proportion of chance agreement P_e is controversial, and it is given in different ways by several authors, according to how they define the concept of agreement by chance (for an extended discussion, see Gwet (2014)).

Fleiss' extension of kappa (Fleiss, 1971), for instance, defines

$$P_e = \sum_{k=1}^{q} p_k^2,$$

where $p_k = \frac{1}{N}\left(\sum_{i=1}^{N} \frac{r_{ik}}{R}\right)$ is the proportion of all assignments which were given to the kth category. In this approach, it is assumed that a random selected rater assigns category k to a random selected target with probability p_k. It follows that under the hypothesis of rater independence, the proportion of chance agreement for category k is p_k^2, and P_e is the total proportion of chance agreement. If the simple assumption is made that $p_k = 1/q$, that is, that each random selected rater assigns category k to a random selected target with the same probability, the kappa index proposed by Brennan and Prediger (1981) is obtained (Marasini et al., 2016 support the utilization of this index and have developed a weighted version for ordinal scales). Light (1971) proposed a slightly different extension that Conger (1980) proved to be equivalent to the average of all pairwise Cohen's kappas for the raters. In this approach, it is assumed that a random selected rater assigns category k to a random selected target according to his personal marginal distribution of the scale. P_e can be computed as the average chance agreements in the contingency tables corresponding to each pair of raters (or by a simplified formula provided in Gwet 2014, p. 53). A conceptually different approach in the definition of the proportion of chance agreement is presented in Gwet (2008), where it is assumed that only an unknown portion of observed ratings is subject to randomness and that chance agreement occurs when at least one rater rates a target randomly. Then, the estimation problem concerns (1) the conditional probability that raters agree given that one of them (or both) has performed a random rating and (2) the probability that one rater (or both) has performed a random rating (see Gwet, 2008 for details of the estimation problem). The proportion of chance agreement is

$$P_e = (\frac{1}{q-1})\sum_{k=1}^{q} p_k(1-p_k),$$

where $p_k = 1/N\left(\sum_{i=1}^{N} \frac{r_{ik}}{R}\right)$ is the proportion of all assignments which were given to the kth category, and the resulting kappa extension is usually named Gwet's AC_1 coefficient. The AC_1 coefficient does not seem to be influenced by the marginal distributions of the categories of the scale observed for each rater. Another kappa extension, frequently applied in the field of communication, was proposed by Krippendorff (1970, 2012). The values assumed in the applications by Krippendorff's kappa extension are very similar to Fleiss' kappa extension, especially when there are no missing data and 5 or more raters.

All the previous kappa extensions presented assume a maximum value equal to 1 in the case of a perfect agreement between the ratings of the raters (in this case, it is $P_a = 1$), a value equal to 0 when the agreement found in the observed ratings is equal to that obtained as a result of the chance ($P_a = P_e$) and negative values if the agreement is less than that obtained as a result of chance ($P_a < P_e$). Although not always coincident, indications have been provided on how to interpret the values that can be assumed; we can say that generally values lower than 0.6 are found in correspondence with low or moderate levels of agreement, values between 0.6 and

0.8 indicate a good level of agreement, and values above 0.8 indicate an excellent level of agreement.

The extensions of kappa have some drawbacks: (1) most of them cannot be computed for only one target ($N = 1$), because in that case the agreement expected by chance is not defined or based on information statistically not relevant; (2) they are formulated in terms of agreement statistics based on all pairs of raters, but some authors argue that simultaneous agreement among three or more raters can be alternatively considered (e.g., see Warrens, 2012); (3) agreement expected by chance depends on the observed proportions of targets allocated to the categories of the scale by each rater, and this implies that the measure of agreement depends on the marginal distributions of the categories of the scale observed for each rater, a source of some paradoxes (for this aspect, see, e.g., Marasini et al., 2016, where a modification of Fleiss' kappa, not affected by this dependence, is proposed). In particular, overall measures of agreement allow us to analyze the agreement between multiple raters for a whole group of targets but not for a single target. When the agreement is poor, it is often necessary to identify the targets with low levels of agreement to request the raters for specific comparisons on the single cases rated, and this can change the scale definition in order to improve the consistency of the rating procedure. In the next section, a proposal of a target-specific measure of agreement by O'Connell and Dobson (1984) is discussed, emphasizing some limitations, and a new measure is proposed based on the homogeneity index to measure the dispersion of a nominal variable (e.g., Leti, 1983).

2 Target-specific Measures of Interrater Agreement for Nominal Scales

In this section, a representation of the observed ratings as a multivariate matrix targets $\times$ raters is assumed to present the index proposed by O'Connell and Dobson (1984). Referring to the response profile in row i of Table 1, a target-specific chance-corrected measure of agreement for several raters using nominal (or ordinal) categories is given by

$$S_i = 1 - D_i/\Delta,$$

where D_i represents the overall disagreement on the whole response profile i ($i = 1, 2, ..., N$) and Δ is the disagreement expected by chance (see O'Connell, & Dobson, 1984, Eq. (6)), computed on the basis of the observed proportion of targets that each rater allocates to the categories of the scale. The computation of the overall disagreement D_i can be explained by the small example in Table 3.

A disagreement function for the response profile i and a pair of raters is defined as 0 if raters agree and 1 if raters disagree (other choices should be considered for ordinal scales). So, the disagreement between rater 1 and rater 2 is 1, the disagreement

Table 3 Judgements of seven raters on a patient (5-category scale)

Patient	Rater 1	Rater 2	Rater 3	Rater 4	Rater 5	Rater 6	Rater 7
i	Categ. 4	Categ. 2	Categ. 4	Categ. 2	Categ. 3	Categ. 4	Categ. 3

between rater 1 and rater 3 is 0, etc. The overall disagreement D_i on the whole response profile i is obtained as the sum of disagreements on all twenty-one possible pairs of different raters. So, the value $D_i = 16$ is obtained, with respect to a maximum value of disagreement that, with 5 categories and 7 raters, is 19. S_i takes the value 1 when there is perfect agreement; it is positive when the agreement is better than chance, and negative otherwise. An overall measure of agreement across subjects S_{av} can be obtained as the arithmetic average of the S_i individual values.

The index S_i shares the same drawbacks of the overall measures of agreement already discussed: (1) it cannot be computed for only one target ($N = 1$) because in that case the disagreement expected by chance Δ is not defined; (2) it is formulated in terms of agreement statistics based on pairs of raters; (3) agreement expected by chance depends on the marginal distributions of the categories of the scale observed for each rater. A different approach is considered below, based on the well-known homogeneity index O to measure the dispersion of a nominal variable having q categories (e.g., Leti, 1983), given by

$$O = \sum_{k=1}^{q} f_k^2,$$

where f_k is the observed proportion of ratings in category k ($k = 1, 2, ..., q$). The index is equal to 1 in the case of maximum homogeneity (perfect agreement), and $1/q$ in the case of maximum heterogeneity (total disagreement, for each category k is $f_k = 1/q$).

O depends on the number of categories, so the normalization in the interval [0,1] is given by

$$O_{rel} = (qO - 1)/(q - 1).$$

It is not difficult to verify that O_{rel} is the normalized variance of the proportions f_k ($k = 1, 2, ..., q$), and it is equal to 1 minus Gini's heterogeneity index (see also Capecchi & Iannario, 2016). In order to apply the normalized homogeneity index O_{rel} to the distribution of the seven raters with respect to the five categories (Table 4), it is convenient to reformulate the index with respect to subject i as

$$O_{rel,i} = \frac{\left(q \sum_{k=1}^{q} \left(\frac{r_{ik}}{R}\right)^2 - 1\right)}{(q - 1)}.$$

Table 4 Distribution of the seven raters with respect to the five categories of the scale in Table 3

Patient	Category 1	Category 2	Category 3	Category 4	Category 5	Total
i	0	2	2	3	0	7

Thus, the target-specific measure of agreement $O_{rel,i}$ is equal to zero for total disagreement and one for perfect agreement. Some experiences with real applications indicate thresholds for the interpretation of $O_{rel,i}$ similar to those presented for kappa in Sect. 2.

For the distribution in Table 4, a value $O_{rel,i} = 0.18$ is obtained, showing a low level of agreement among the seven ratings.

An overall measure of agreement on a whole group $\overline{O}_{rel}$ can be easily obtained as the arithmetic average of the individual values $O_{rel,i}$ as

$$\overline{O}_{rel} = \frac{1}{N}\sum_{i=1}^{N} O_{rel,i}.$$

Advantages of the new measures proposed are that (1) they can be computed for only one target ($O_{rel,i}$), so they can be applied also when each target is rated by a different set of raters and/or the number of raters is different for each target; (2) they are not formulated in terms of agreement statistics based on pairs of raters but in terms of statistical variability of the distributions of the raters with respect to the categories of the scale; (3) they do not depend on the marginal distributions of the categories of the scale observed for each rater, so they are not affected by paradoxes like many of the kappa extensions previously presented. It can be shown that in this particular case the number of raters is equal to the number of categories of the scale ($R = q$); $\overline{O}_{rel}$ is equal to P_a. Berry and Mielke (1988) presented seven basic attributes that a measure of agreement should embody; some of them apply to the case of nominal scales considered in this paper. Namely, $\overline{O}_{rel}$ is constructed by an approach based on statistical variability and cannot be considered a chance-corrected measure of agreement. In our opinion, most of the assumptions that allow the computation of the proportion of chance agreement P_e are unrealistic and produce paradoxes. In line with Gwet (2008), we think that in most applications raters do not assign all ratings by chance, and only a portion of raters assign by chance to a small portion of targets. From this point of view, low values of the target-specific index $O_{rel,i}$ can be an indication of targets for which ratings could be affected by chance assignments and for which the group of raters could be called for a revision. Moreover, as previously remarked, an approach based on statistical variability allows us to avoid some paradoxes and limitations by which some of the kappa extensions are affected (exceptions seem to be Gwet's AC_1 and the extension of kappa proposed in Brennan and Prediger (1981); see Quarfoot and Levin (2016)).

Berry and Mielke (1988) advocate that a measure of agreement should be able to analyze multivariate data and evaluate information from more than two observers;

both these requirements are satisfied by $O_{rel,i}$ and $\overline{O}_{rel}$. On the other hand, the new approach considered is mainly descriptive, so $O_{rel,i}$ and $\overline{O}_{rel}$ lack a statistical base, which should be considered in future developments of this study.

3 Application

Data considered are about a study with seven nosologists assessing the cause of death of 35 hypertensive patients by using their death certificates (Woolson, 1987). The scores were assigned by the following categories: 1=Arteriosclerotic disease, 2=cerebrovascular disease, 3=other heart disease, 4=renal disease, and 5=other disease. The marginal proportions of ratings for the five categories were 0.21, 0.17, 0.19, 0.27, and 0.16, respectively. Some results are presented for the method based on the O_{rel} index.

The subjective values of $O_{rel,i}$ allowed the detection of low level of agreement for many evaluations (28.6% of the $O_{rel,i}$ values less than 0.4), which calls for a possible revision of the assessment procedure. It can be also interesting to analyze some descriptive statistics provided in Table 5 for the comparison of S_i and $O_{rel,i}$. The mean values for global agreement are $S_{av} = 0.48$ and $\overline{O}_{rel} = 0.56$. S_i values show higher dispersion with respect to $O_{rel,i}$ values. The measures are almost perfectly correlated ($r = 0.99$).

We also computed the values of other indices of interrater agreement: the values of the average Cohen's kappa (Light's extension of kappa), Fleiss kappa, Krippendorff kappa, and Gwet's AC_1 were approximately equal to 0.48 and coincide with S_{av}; the value of Brennan and Prediger kappa was 0.40. So, all indices show a moderate level of agreement between the seven nosologists. It is interesting to point out that if we increase the level of agreement between raters by collapsing the five categories into the two strongly unbalanced categories cerebrovascular disease (marginal proportion 0.17) and all other diseases (marginal proportion 0.83), the values of S_{av}, Light kappa, Fleiss kappa, and Krippendorff kappa remain almost the same, while the new values of $\overline{O}_{rel}$, Gwet's AC_1, and Brennan and Prediger kappa increase to 0.75, 0.80, and 0.71, respectively, according to the new high level of agreement determined by the aggregation. It is not uncommon in applications to have highly unbalanced categories; this happens, for example, when a diagnostic category is rare or when for some reason the raters use almost exclusively very few levels of the scale. The

Table 5 Some descriptive statistics for S_i and $O_{rel,i}$ values

	N	Mean	Std. Dev.	CV
S_i	35	0.48	0.27	56.5
$O_{rel,i}$	35	0.56	0.23	42.1

possibility to compute also the target-specific values $O_{rel,i}$ is an advantage of $O_{rel,i}$ with respect to the indices Gwet's AC_1 and Brennan and Prediger kappa.

4 Conclusion

Generalizations of Cohen's kappa to analyze the agreement on a nominal scale for the case of more than two raters were reviewed emphasizing some limitations frequently encountered in their application. These indices are used to analyze the agreement between multiple raters for a whole group of targets. Measures of agreement on a single target have been considered less frequently, in spite of the fact that having evaluations of agreement on a single case is particularly useful. O'Connell and Dobson (1984) proposed a single-target measure of interrater agreement that, however, depends on knowledge of the whole group of ratings and on the marginal distributions of the categories of the scale observed for each rater. A descriptive approach to the analysis of absolute interrater agreement has been proposed, formulated in terms of statistical variability of the distributions of the raters with respect to the categories of the scale, rather than in terms of agreement statistics based on pairs of raters. A target-specific measure and a global measure of agreement are proposed that present some advantages with respect to the kappa extensions more frequently encountered in application. The indices proposed are mainly considered as measures of the size of interrater agreement, therefore future developments of this study may concern the definition of reliable thresholds useful in application. Besides, the sampling properties of $\overline{O}_{rel}$ should be studied and compared with competitor's measures like Gwet's AC_1 coefficient. Finally, we notice that a measure of interrater agreement for ordinal data recently proposed and applied in educational studies follows an approach similar to the present proposal (Bove et al., 2021) where a measure of dispersion for ordinal variables is considered instead of the homogeneity index.

References

Berry, K. J., & Mielke, P. W. (1988). A measure of interrater absolute agreement for ordinal categorical data. *Educational and Psychological Measurement, 48*, 921–933.

Bove, G., Conti, P. L., & Marella, D. (2021). A measure of interrater absolute agreement for ordinal categorical data. *Statistical Methods and Applications, 30*(3), 927–945.

Brennan, R. L., & Prediger, D. J. (1981). Coefficient Kappa: some uses, misuses, and alternatives. *Educational and Psychological Measurement, 41*, 687–699.

Capecchi, S., & Iannario, M. (2016). Gini heterogeneity index for detecting uncertainty in ordinal data surveys. *Metron, 74*, 223–232.

Cohen, J. (1960). A coefficient of agreement for nominal scales. *Educational and Psychological Measurement, 20*, 213–220.

Conger, A. J. (1980). Integration and generalization of kappas for multiple raters. *Psychological Bulletin, 88*, 322–328.

Fleiss, J. L. (1971). Measuring nominal scale agreement among many raters. *Psychological Bulletin, 76*, 378–382.
Gwet, K. L. (2008). Computing inter-rater reliability and its variance in the presence of high agreement. *British Journal of Mathematical and Statistical Psychology, 61*, 29–48.
Gwet, K. L. (2014). *Handbook of inter-rater reliability* (4th ed.). Gaithersburg, MD, USA: Advanced Analytics, LLC.
Krippendorff, K. (1970). Estimating the reliability, systematic error and random error of interval data. *Educational and Psychological Measurement, 30*, 61–70.
Krippendorff, K. (2012). *Content Analysis: An Introduction to its Methodology*, Thousand Oaks. CA: Sage Publications.
Landis, J. R., & Koch, G. G. (1977). An application of hierarchical Kappa-type statistics in the assessment of majority agreement among multiple observers. *Biometrics, 33*, 363–374.
Leti, G. (1983). *Statistica descrittiva*. Bologna: Il Mulino.
Light, R. J. (1971). Measures of response agreement for qualitative data: some generalizations and alternatives. *Psychological Bulletin, 76*, 365–377.
Marasini, D., Quatto, P., & Ripamonti, E. (2016). Assessing the interrater agreement for ordinal data through weighted indexes. *Statistical Methods in Medical Research, 25*, 2611–2633.
O'Connell, D. L., & Dobson, A. J. (1984). General observer-agreement measures on individual subjects and groups of subjects. *Biometrics, 40*, 973–983.
Quarfoot, D., & Levin, R. A. (2016). How robust are multirater interrater reliability indices to changes in frequency distribution? *American Statistician, 70*(4), 373–384.
Warrens, M. J. (2012). Equivalences of weighted kappas for multiple raters. *Statistical Methodology, 9*, 407–422.
Woolson, R. F. (1987). *Statistical methods for the analysis of biomedical data*. New York: Wiley.

Hierarchical Clustering of Income Data Based on Share Densities

Francesca Condino

Abstract Starting from a situation where a reference population of income earners is naturally divided into sub-groups, the aim of this work is to explore the similarity of the sub-populations in terms of income inequality. To this end, we propose to extract information from a particular function, the so-called share density, strongly related to different inequality measures, such as the Gini and Theil indexes. The Jensen-Shannon dissimilarity measure is proposed to evaluate the discrepancy across share densities, and a hierarchical clustering algorithm is employed to find the family of partitions. Results regarding data from the Survey on Households Income and Wealth (SHIW) by Bank of Italy are shown.

Keywords Tail inequality · Dissimilarity measure · Income concentration

1 Introduction

The analysis of income inequality is a central theme in the economic and social debate, as proven by the wide existing literature on this argument. A great variety of methodologies have been proposed, not only to quantify the concentration level in a population but also to compare different populations, geographical areas, groups and so on. Most of these methodologies are based on the Lorenz curve, proposed by Lorenz (1905) and currently still widely used. Starting from the Lorenz curve, different inequality measures and tools have been proposed to explore the disparity in a population and to construct different methodologies for group comparisons. With reference to this second issue, Fields and Fei (1978) identify two approaches, namely the cardinal and the ordinal approach, according to whether the comparison is implemented by means of summary measures or on the basis of the Lorenz domination criterion. Actually, when the comparison across groups of income earners is

F. Condino (✉)
Department of Economics Statistics and Finance Giovanni Anania, University of Calabria, Arcavacata, Italy
e-mail: francesca.condino@unical.it

L. Grilli et al. (eds.), *Statistical Models and Methods for Data Science*, Studies in Classification, Data Analysis, and Knowledge Organization,
https://doi.org/10.1007/978-3-031-30164-3_3

limited to the comparison between synthetic inequality index values, such as Gini or Theil, the obtained results can be very inaccurate, since similar values of these measures can hide very different structures in income distribution. On the other hand, checking the Lorenz dominance can lead to unclear comparisons, since the curves may intersect (Davies & Hoy, 1995). The attempt of this work is to develop a procedure for quantifying the dissimilarity across sub-groups on the basis of the whole structure of inequality, rather than of specific indicators. In doing so, we consider a dissimilarity measure borrowed from information theory, a context having many contact points with inequality field, as proven by the large body of literature that has its roots in the work of Theil (1967) and its successive developments in Cowell (1980a, 1980b), Shorrocks (1980) and Cowell and Kuga (1981), to cite a few. In this context, more recently, Rohde (2016) proposes the use of a symmetric entropy statistic, the J-divergence measure, to study income inequality.

In order to consider the whole structure of inequality, the chosen dissimilarity measure will be applied to the so-called share densities, some particular functions obtained from the Lorenz curve and related to the probability that a unit of amount, chosen at random, is earned by a specific percentile range of the population. Finally, this dissimilarity will be used for clustering purposes.

2 Lorenz Curve and Share Density: A Parametric Approach

Although many methods to estimate income distribution and its properties are available, here we focus on the parametric approach, which requires the choice of an a priori functional form for income data description. Given a parametric model, it is possible to obtain the Lorenz curve and its properties. To this end, let Y be the continuous random variable (rv) describing income and $F(y; \boldsymbol{\theta})$ its distribution function (df), where $\boldsymbol{\theta}$ is the parameter vector. It is well known that, starting from the expression of df, the parametric Lorenz curve $L(u, \theta)$ can be obtained as

$$L(u; \theta) = \frac{\int_0^u F^{-1}(t; \theta) dt}{\mu}$$

where $F^{-1}(t; \theta) = \sup\{y|F(y) \leq t\}$, for $t \in [0, 1]$ is the quantile function and $\mu = \int_Y y dF(y; \theta)$ is the mean of Y.

In economic literature, the Lorenz curve is a well-known and widely used tool for analysing income inequality. Since its proposal, in 1905 (Lorenz, 1905), a lot of investigation has been suggested among statisticians and economists, generating a fertile field of study. It is known that each Lorenz curve $L(u; \theta)$ ($u \in [0, 1]$) can be viewed as a distribution function on the unit interval and it is possible to consider its derivative, $l(u; \theta) = \frac{\partial L(u;\theta)}{\partial u}$, as a parametric density function.

This function was rarely mentioned and used to explore inequality, although it furnishes different information about it, as suggested by Rohde (2008), who has shown that the two well-known Theil's inequality indexes, T_L and T_T, can be directly obtained from $l(u)$. In particular, Theil's T_T index coincides with the Shannon entropy of $l(u)$, changed in sign. Some reference to the Lorenz density can be found in Farris (2010), where this curve is referred to as share density. Afterwards, the concept of Lorenz density is recalled in Zizler (2014), Kämpke and Radermacher (2015) and Shao (2021). It is worth to note that comparing the Gini or Theil index values between groups is equivalent to comparing specific characteristics of their corresponding share densities. Therefore, here we want to quantify the discrepancy among groups of income earners by quantifying the dissimilarity between the whole share densities. To do this, the choice of a proper measure is required.

3 Hierarchical Algorithm Based on JS Dissimilarity

Let us consider a reference population naturally divided in a set E of K groups of income earners, i.e. $E = \{\omega_1, ..., \omega_K\}$, each characterized by its own share density function $l_1, ..., l_K$. In this paper, we consider an agglomerative clustering algorithm, starting with K distinct non-overlapping clusters $(C_1, ..., C_K)$ each containing a single object. The algorithm is quite similar to that used for classical data structure of the type $units \times variables$, and the main difference regards the calculation of the dissimilarity matrix, which must take into account the particular nature of the objects to be partitioned and their descriptors, namely the share densities. In order to analyse the existing differences among various groups of income earners, this dissimilarity measure will be considered in connection with the Lorenz density. Therefore, let $F_1, ..., F_K$ (for brevity of notation, the arguments of the functions are omitted) be the parametric df corresponding to K different groups of income earners. As previously mentioned, we can obtain the corresponding Lorenz curve $L_1, ..., L_K$ and the corresponding derivatives with respect to $u, l_1, ..., l_K$. With the aim to quantify the discrepancy between each couple of share densities, of all possible measures of dissimilarity among density functions, in this work we consider the Jensen-Shannon (JS) dissimilarity, also called total divergence to the average. This is an information theoretic measure, and, therefore, strongly related to the entropy-based measures of inequality, originally introduced by Theil (1967). Hence, given two share densities, l_k and $l_{k'}$, the JS divergence is given by

$$D_{JS}(l_k, l_{k'}) = H(l_m) - \pi_k H(l_k) - \pi_{k'} H(l_{k'}) \tag{1}$$

where $l_m = \pi_k \cdot l_k + \pi_{k'} \cdot l_{k'}$ is their mixture. The weights π_k and $\pi_{k'}$, as suggested by Bishop et al. (2003), represent the income shares, respectively, for the k-th and k'-th group. The JS dissimilarity can be alternatively written as

$$D_{JS}(l_k, l_{k'}) = \pi_k \cdot D_{KL}(l_k||l_m) + \pi_{k'} \cdot D_{KL}(l_{k'}||l_m) \tag{2}$$

where $D_{KL}\left(l_j||l_m\right) = \int_0^1 l_j\left(u\right)\log\frac{l_j(u)}{l_m(u)}du$, for $j = k, k'$, is the so-called Kullback-Leibler divergence between each share density and the mixture l_m.

From (1), it can be noted that the JS divergence directly depends on Theil's T_T index for the two groups, since the entropies of share densities are involved; moreover, it depends on Theil's T_T index for the mixture of the two sub-populations, changed in sign. Therefore, it represents the difference between Theil's inequality within the two groups and Theil's inequality under the hypothesis that individuals belong to the population made up of the two mixture components. From (2), it is also evident that this measure takes into account the discrepancy of each function from the mixture, along the whole unit interval, so that it will be influenced by any existing deviation that may occur in some segment of the examined population. As a consequence, the clustering procedures based on the JS divergence will exploit these discrepancies.

In this paper, we consider the hierarchical clustering procedure that seeks to progressively merge groups of income earners having the smallest JS dissimilarity in terms of share density functions. Therefore, the starting point is represented by the dissimilarity matrix $\boldsymbol{D} = \{d_{JS}(l_k, l_{k'})\}$, whose elements are the JS dissimilarities between each pair of share densities, computed through (1), for $k \neq k' = 1, ..., K$. From the initial partition $P^{(1)} = (C_1, ..., C_K)$, where C_k, for $k = 1, ..., K$, contains one observation ω_k; at each step s, the two closest clusters in $P^{(s)}$, i.e. clusters for which $d(C_k, C_{k'})$ in $\boldsymbol{D}^{(s)}$ is the smallest, are merged, until the final partition $P^{(K)}$, made up of a single cluster containing all the objects $\omega_1, ..., \omega_K$, is obtained. As for the classical approach, different criteria can be used to compute the distances $d(C_k, C_{k'})$ between clusters and update the partition.

4 An Application

In this section, data from the Survey on Households Income and Wealth (SHIW), carried out by Bank of Italy in 2016, are considered. To take into account the composition of households, equivalent incomes (in ten of thousands) are obtained, using the OECD-modified equivalent scale.

For this application, we consider the Dagum model (Dagum, 1977, 1980) to fit income distribution. Indeed, the Dagum model has been widely used to study income and wealth, thanks to its ability in reproducing particular features of these kinds of data. Kleiber and Kotz (2003) and Kleiber (2008) provided an exhaustive description of the genesis of this model and a comprehensive review of its characteristics.

Here, we remember that the Dagum *rv* has a positive support and its *df* is given by $F_{Da}(y; \beta, \lambda, \delta) = \left(1 + \lambda y^{-\delta}\right)^{-\beta}$, where all the involved parameters are positive. In particular, λ is a scale parameter and β and δ are shape parameters.

In order to obtain the estimates for each region, we consider the maximum likelihood estimation method and income data at the regional level. Having obtained the estimates $\hat{\beta}_k, \hat{\lambda}_k, \hat{\delta}_k$, for $k = 1, ..., 20$, it is possible to compute some synthetic values for describing various population features, such as average income, $\hat{\mu}_k$ and

Table 1 Fitted means, negative entropy and Gini index for Italian regions

Regions	$\hat{\mu}_k$	$-\hat{H}(l_k)$	$\hat{G}_k$
Piedmont	2.0797	0.1277	0.2751
Aosta Valley	2.3134	0.1195	0.2663
Veneto	1.9739	0.1244	0.2706
Friuli	2.3140	0.1226	0.2693
Emilia Romagna	2.3971	0.1141	0.2602
Tuscany	2.3648	0.1131	0.2588
Abruzzo	1.9542	0.1237	0.2717
Calabria	1.3472	0.1425	0.2910
Sardinia	1.5772	0.1361	0.2843
Lombardy	2.4798	0.1632	0.3064
Molise	1.7789	0.1873	0.3294
Campania	1.3461	0.1815	0.3252
Apulia	1.4558	0.1557	0.3026
Basilicata	1.5191	0.1850	0.3287
Sicily	1.4610	0.1633	0.3071
Trentino	2.2247	0.1008	0.2408
Liguria	2.2482	0.1356	0.2782
Umbria	1.9897	0.1024	0.2456
Marche	2.1809	0.1053	0.2475
Lazio	1.9972	0.1437	0.2883

Gini inequality index $\hat{G}_k$. Moreover, for this model it is easy to obtain the derivative of Lorenz curve and, consequently, through the numerical integration method, its entropy. All these results are reported in Table 1.

For each couple of regions, the mixture of the corresponding Lorenz densities is considered, and the elements of the JS dissimilarity matrix $\boldsymbol{D}^{(1)}$ are computed to initialize the above-described algorithm. Table 2 reports the obtained values, in thousandths, from which it is evident that the minimum value is obtained in correspondence of the couple {Aosta Valley, Emilia Romagna}. Therefore, these two regions are merged in the first step of the algorithm. On the other hand, the maximum value for JS dissimilarity is obtained in correspondence of the couple {Trentino, Campania}.

To compute the distances between clusters, the unweighted pair group method with averaging (UPGA) is chosen, after comparing its performance with that of different agglomeration methods based on dissimilarity matrix $\boldsymbol{D}$. In particular, the value of the cophenetic correlation coefficient from the UPGA method (0.6286) results higher than that obtained from the weighted pair group method with averaging (WPGA; 0.5920), the complete (0.6176) and the single (0.5127) method, suggesting a more faithful representation of the original pairwise dissimilarities among regions.

Table 2 JS distance matrix elements for Italian regions (in thousandths)

	Piedmont	Aosta Valley	Lombardy	Trentino	Veneto	Friuli	Liguria	E. Romagna	Tuscany	Umbria
Aosta Valley	0.083									
Lombardy	0.865	0.342								
Trentino	0.973	0.478	1.477							
Veneto	0.144	0.106	0.761	0.724						
Friuli	0.095	0.082	0.642	0.880	0.104					
Liguria	0.615	0.400	0.513	0.777	0.425	0.587				
E. Romagna	0.178	0.069	1.335	0.604	0.157	0.097	0.668			
Tuscany	0.213	0.077	1.317	0.527	0.155	0.110	0.623	0.073		
Umbria	0.503	0.215	1.283	0.262	0.368	0.399	0.776	0.220	0.182	
Marche	0.709	0.245	1.452	0.158	0.451	0.484	0.604	0.390	0.320	0.133
Lazio	0.435	0.276	0.329	1.136	0.348	0.411	0.280	0.662	0.651	0.913
Abruzzo	0.163	0.143	0.891	1.710	0.371	0.278	1.311	0.216	0.282	0.873
Molise	0.731	1.838	0.302	3.126	1.040	1.644	1.189	1.058	1.146	2.706
Campania	1.319	0.670	0.455	3.576	1.586	1.462	1.557	2.072	2.176	2.972
Apulia	0.414	0.361	0.341	2.322	0.604	0.615	0.900	0.827	0.901	1.696

(continued)

Table 2 (continued)

	Piedmont	Aosta Valley	Lombardy	Trentino	Veneto	Friuli	Liguria	E. Romagna	Tuscany	Umbria
Basilicata	0.680	1.738	0.334	3.241	1.021	1.573	1.310	1.012	1.108	2.706
Calabria	0.168	0.321	0.461	1.972	0.367	0.388	0.993	0.347	0.413	1.239
Sicily	0.774	0.480	0.210	2.165	0.821	0.860	0.673	1.247	1.280	1.822
Sardinia	0.123	0.160	0.604	1.761	0.296	0.242	0.946	0.310	0.374	1.027
	Marche	Lazio	Abruzzo	Molise	Campania	Apulia	Basilicata	Calabria	Sicily	
Lazio	0.899									
Abruzzo	1.070	0.832								
Molise	2.113	0.623	1.907							
Campania	3.146	0.940	1.546	0.198						
Apulia	1.755	0.411	0.706	0.417	0.409					
Basilicata	2.162	0.668	1.677	0.232	0.185	0.366				
Calabria	1.218	0.490	0.254	1.079	0.666	0.232	0.888			
Sicily	1.818	0.346	1.222	0.354	0.427	0.308	0.399	0.578		
Sardinia	1.161	0.507	0.184	1.045	0.983	0.307	0.910	0.097	0.706	

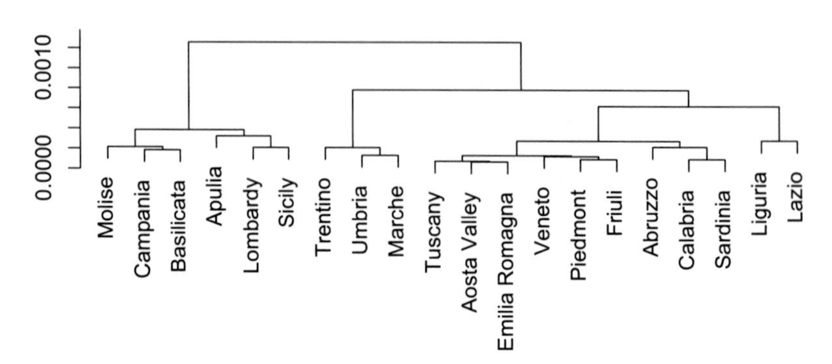

Fig. 1 Dendrogram based on UPGA agglomeration method

Therefore, according the UPGA agglomeration method, given two clusters C_j and $C_{j'}$, consisting of N_j and $N_{j'}$ regions, respectively, and belonging to the partition $P^{(s)}$, the distance between them is

$$d(C_j, C_{j'}) = \frac{1}{N_j \cdot N_{j'}} \sum_{\omega_k \in C_j} \sum_{\omega'_k \in C'_j} \{d_{JS}(l_k, l_{k'})\}$$

where l_k is the share density for the region $\omega_k \in C_j$ and $l_{k'}$ is the share density for the region $\omega_{k'} \in C_{j'}$, $j \neq j'$.

The dendrogram in Fig. 1 suggests the presence of three or four clusters. In particular, by considering a partition made up of three clusters, we obtain the following groups:
$C_1 = \{$*Abbruzzo, Piedmont, Sardinia, Friuli, Emilia Romagna, Aosta Valley, Calabria, Veneto, Tuscany, Liguria, Lazio*$\}$
$C_2 = \{$*Molise, Campania, Basilicata, Sicily, Lombardy, Apulia*$\}$

$C_3 = \{$*Trentino, Marche, Umbria*$\}$.

As we can see from the results reported in Table 3, clusters seem clearly characterized, with regions having generally lower concentration of income belonging to clusters 1 and 3, and regions with higher concentration levels included in cluster 2. Moreover, the element that mainly differentiates cluster 1 from cluster 3 seems to be the different concentration level in correspondence to the poorest households. Indeed, if we look at the Gini index for the poorest households ($G^{(L)}$), i.e. households having income lower than the median, and the Gini index for the richest $G^{(L)}$, having income higher than the median, we note that the regions belonging to cluster 1 tends to show higher inequality in correspondence to the left tail than regions belonging to cluster 3. This phenomenon is less evident in the Liguria region, as confirmed by the silhouette plot in Fig. 2 and, when the number of clusters equal to four are considered,

Table 3 Observed Gini index for total households (G_k), for poorer ($G_k^{(L)}$) and richer earners ($G_k^{(U)}$) and membership cluster for Italian regions

Regions	G_k	$G_k^{(L)}$	$G_k^{(U)}$	Cluster
Piedmont	0.2756	0.1854	0.1717	1
Aosta Valley	0.2594	0.1967	0.1627	1
Veneto	0.2848	0.1818	0.1937	1
Friuli	0.2699	0.1860	0.1593	1
Emilia Romagna	0.2601	0.1761	0.1616	1
Tuscany	0.2562	0.1631	0.1602	1
Abruzzo	0.2589	0.1764	0.1405	1
Calabria	0.2743	0.1982	0.1651	1
Sardinia	0.2851	0.2056	0.1631	1
Liguria	0.2845	0.1572	0.2032	1
Lazio	0.2878	0.1830	0.1840	1
Lombardy	0.3096	0.1804	0.2249	2
Molise	0.3167	0.1663	0.2133	2
Campania	0.3216	0.2210	0.2047	2
Apulia	0.2922	0.1879	0.1848	2
Basilicata	0.3273	0.2242	0.1938	2
Sicily	0.3064	0.1965	0.2027	2
Trentino	0.2758	0.1383	0.2315	3
Umbria	0.2451	0.1555	0.1503	3
Marche	0.2415	0.1426	0.1465	3

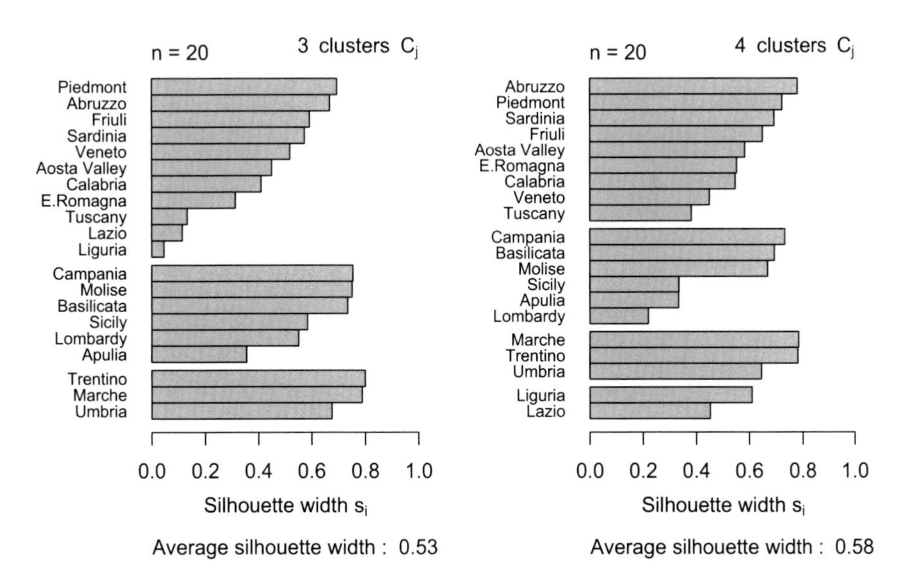

Fig. 2 Dendrogram and silhouette plot

it constitutes a new cluster together with Lazio, while the other two clusters remain unchanged. In this case, the average silhouette increases from 0.53 to 0.58.

5 Conclusion

In this paper, a non-conventional hierarchical clustering procedure is proposed with the aim to group sub-populations of income earners having a similar structure in terms of inequality. In this context, the objects to be partitioned are described by means of the derivative of Lorenz curve, the so-called share density. As shown by different authors, this function furnishes various information regarding inequality. Therefore, a measure of the discrepancy across density functions, strongly related to information theory, namely the JS dissimilarity, is considered to quantify the existing differences among inequality structure of different sub-populations. The application to real data shows the potentiality of this proposal. A parametric approach based on the Dagum model is considered to fit income data, and the obtained findings suggest that this method allows us to properly partition the share densities on the basis of their global and local behaviour. Indeed, through this approach it is possible to take into account inequality in different segments of the population and to explore more in depth the different structures of inequality across groups.

References

Billard, L., & Diday, E. (2020). *Clustering methodology for symbolic data*. Hoboken, NJ: Wiley.

Bishop, J. A., Chow, K. V., & Zeager, L. A. (2003). Decomposing Lorenz and concentration curves. *International Economic Review, 44*, 965–978.

Chotikapanich, D., Griffiths, W. E., Hajargasht, G., Karunarathne, W., & Rao, D. S. P. (2018). Using the GB2 income distribution. *Econometrics, 6*, 21.

Cowell, F. A. (1980a). Generalized entropy and the measurement of distributional change. *European Economic Review, 13*, 147–159.

Cowell, F. A. (1980b). On the structure of additive inequality measures. *The Review of Economic Studies, 47*, 521–531.

Cowell, F. A., & Kuga, K. (1981). Additivity and the entropy concept: An axiomatic approach to inequality measurement. *Journal of Economic Theory, 25*, 131–143.

Dagum, C. (1977). A new model of personal distribution: Specification and estimation. *Economie Appliquée, 30*, 413–437.

Dagum, C. (1980). The generation and distribution of income, the Lorenz curve and the Gini ratio. *Economie Appliquée, 33*, 327–367.

Dancelli, L. (1989). Confronti fra le curve di concentrazione Z(p) e L(p) nel modello di Dagum. *Statistica Applicata, 1*, 399–415.

Davies, J., & Hoy, M. (1995). Making inequality comparisons when Lorenz curves intersect. *The American Economic Review, 85*(4), 980–986.

Farris, F. A. (2010). The Gini index and measures of inequality. *American Mathematical Monthly, 117*(10), 851–864.

Fields, G. S., & Fei, J. C. H. (1978). On inequality comparisons. *Econometrica, 46*(2), 303–316.

Kämpke, T., & Radermacher, F. (2015). *Income modeling and balancing: A rigorous treatment of distribution patterns*. Switzerland: Springer.
Kleiber, C., & Kotz, S. (2003). *Statistical size distributions in economics and actuarial sciences*. Wiley.
Kleiber, C. (2008). *The Lorenz curve in economics and econometrics*. London: Routledge.
Lorenz, M. O. (1905). Methods of measuring the concentration of wealth. *Publications of the American Statistical Association, 9*(70), 209–219.
Rohde, N. (2008). Lorenz Curves and generalised entropy inequality measures. In D. Chotikapanich (Eds.), *Modeling Income Distributions and Lorenz Curve. Economic Studies in Equality, Social Exclusion and Well-Being* (Vol. 5). New York: Springer.
Rohde, N. (2016). J-divergence measurements of economic inequality. *Journal of the Royal Statistical Society. Series A (Statistics in Society), 179*(3), 847–870.
Shao, B. (2021). Decomposition of the Gini index by income source for aggregated data and its applications. *Computational statistics*, Epub ahead of print: 1–25.
Shorrocks, A. F. (1980). The class of additively decomposable inequality measures. *Econometrica, 48*, 613–625.
Theil, H. (1967). *Economics and information theory*. Amsterdam: North Holland.
Zizler, P. (2014). Gini indices and the moments of the share density function. *Applications of Mathematics, 59*, 167–175.

Optimal Coding of High-Cardinality Categorical Data in Machine Learning

Agostino Di Ciaccio

Abstract Analyzing categorical data in machine learning generally requires a coding strategy. This problem is common to multivariate statistical techniques, and several approaches have been suggested in the literature. This article proposes a method for analyzing categorical variables with neural networks. Both a supervised and unsupervised approaches were considered, in which the variables can have high cardinality. Some simulated data applications illustrate the interest in the proposal.

Keywords Encoding categorical data · Neural networks · High-cardinality attributes · Optimal scaling

1 Introduction

Most machine learning algorithms cannot be applied directly to categorical data that are generally non-numeric. Their application therefore requires some form of encoding that transforms the categorical features into one or more numeric variables. This problem can pose a serious difficulty if the variables have many categories, a common situation for big data generally including mixed measurement levels of the variables.

We could have categorical variables with tens or hundreds of categories, e.g., names of provinces, postcodes, ATECO codes, list of company names, list of product names, and so on. A further complication is that some categories that we might observe in the future were not observed in the training data. To analyze qualitative variables together with quantitative variables, we can create for each categorical variable an embedding of the categories, in a space with one or more dimensions.

A. Di Ciaccio (✉)
University of Rome La Sapienza, Rome, Italy
e-mail: agostino.diciaccio@uniroma1.it

L. Grilli et al. (eds.), *Statistical Models and Methods for Data Science*, Studies in Classification, Data Analysis, and Knowledge Organization,
https://doi.org/10.1007/978-3-031-30164-3_4

The embedding of strings in a multidimensional space is a well-established procedure in Natural Language Processing (NLP) (Bengio et al., 2003) and is the inspiration for this paper.

Many encoding systems for supervised and unsupervised models have been proposed in the literature, but they have several weaknesses. In a previous paper (Di Ciaccio, 2020), a technique was proposed, the Low Embedding Encoder (LEE), which extends the concept of quantification using an approach similar to word embedding in Natural Language Processing. The aim of LEE was to analyze categorical variables with high cardinality using Neural Networks in a supervised approach. In this paper, this technique is extended to consider also an unsupervised approach, pointing out some relationship over consolidated techniques such as Correspondence Analysis Benzécri (1973) or Optimal Scaling Gifi (1990). In par. 2, there is a review of the techniques proposed in the literature both in a supervised and unsupervised approaches, showing some weaknesses of each technique, with reference to a Machine Learning context. In par. 3, some approaches based on dummy variables are analyzed, then in the successive paragraphs we show a non-linear extension using neural networks, both in the supervised and unsupervised cases. Some examples highlight the interest in our proposal.

2 Quantify Categorical Features: A Review of Existing Methods

In the literature, many methods have been proposed to encode categorical variables (a recent review is Hancock et al., 2020). The classification of the encoding methods can consider several elements: how the method takes into account the variable to predict (the target, if available), how the coding is influenced by the other explanatory variables, how many new variables are created, and what the aim of the analysis is. Several encoding methods for ML have been proposed in the literature (see, for example, Potdar et al., 2017) mainly in a supervised approach. The scikit-learn software library (https://scikit-learn.org/) allows to apply 17 different methods. Here, we will consider the most used techniques, without pretending to be exhaustive, distinguishing them into three groups: encoders that do not use the target or other data; encoders requiring only the target; encoders that use dummy variables.

2.1 Methods that Do Not Consider the Target or the Other Variables

These methods assign the encoding to the categorical variable according to a criterion that does not involve the target or other data. In this way, there is no risk of overfitting, but the encodings obtained are extraneous to the dataset, and the result largely depends on the researcher's choices or is essentially a random quantification and therefore not interpretable. A crude way is to assign integer values to the categories (label encoding), possibly respecting the natural category order (ordinal encoding). The result is a single numerical variable. Note that all the possible categories must be known before assigning the quantifications (integers).

The hash encoder uses a hash function to convert the K categories of a variable to $s << K$ columns, producing a smaller number of dimensions than One-Hot Encoding. Using this method, there is no way to do a reverse lookup and determine what the input was. By this encoder, a large number of categories are depicted in lesser dimensions using a simple function (hash function). This function, which maps the value of a category into one or more numbers, can be extremely efficient as the hash function does not require a predefined dictionary of possible categories and can also code categories not observed in the training step. Hence, multiple values could be represented by the same hash value; this is known as a collision. By default, the Hash encoder uses the md5 hashing algorithm (Yong-Xia et al., 2010). This encoder is sometimes used in big data when the cardinality of a variable is very high.

2.2 *Encoders Requiring only the Target*

There are a lot of methods that use the target to obtain a numerical coding of the categories, while the other explanatory variables do not influence the coding. The result is a single numeric variable for regression tasks, allowing to keep the dimensionality of our data the same as the original data, or multiple numerical variables for classification tasks. Applying target-based encoding often produces data leakage, leading to overfitting and poor predictive performance. To work fine, this method needs a large amount of data and low cardinality of categorical features. To overcome data leakage, it was proposed to add noise, or use cross-validation technics or other forms of regularization.

Simple Target Encoder is a popular method for regression tasks. The method consists in assigning the conditional mean target value to each category of the explanatory variable. For classification tasks, where also the target is categorical, we encode the explanatory categorical variable with $m - 1$ new features (m is the number of classes of the target) which indicate the probability of a data point belonging to the classes. These probabilities are estimated from the conditional frequencies, if we have not set any priors. Leave One Out Encoder tries to limit data leakage excluding the current row's target when calculating the mean target for a category. In this way, the same category can be coded differently on different units in the training set, while, on new data, the mean of the category quantifications obtained is used. The K-fold Target Encoding works in a similar way. CatBoost Encoder is like leave-one-out encoding but calculates the values "on the fly" to avoid overfitting. In fact, to calculate the quantification for the j-th observation we can only use observations with a lower position in the dataset. This procedure is repeated several times on shuffled versions of the dataset, and then the results are averaged. Other methods in this approach are based on the "contrast" between one category and the other categories of the variable; they are called contrast encoders. For example, the Helmet encoder is commonly used for regression tasks. It compares the target mean of each category of the variable with the target mean of the subsequent levels of the variable. This type of coding requires ordered categorical variables.

2.3 One-Hot Encoding (OHE)

The previous approaches do not take into account the other explanatory variables or the model used, but possibly only the target, and the quantifications are fixed before the model is estimated and therefore are not model parameters. In OHE, a new binary variable is introduced for each category, indicating the presence or absence of the category on the observed units. Some methods exclude the dummy for one category (named the reference level) due to the multicollinearity problem (the dummy variable trap) but applying machine learning models it is necessary to include all the categories, otherwise we would never consider the omitted category. OHE is frequently applied in statistical analysis, e.g., in the linear regression model. Assuming a least square loss, considering only one categorical explanatory variable coded as dummy variables and one quantitative dependent variable, the optimal coefficient for each dummy is the mean of the dependent variable for the corresponding category minus the mean of the dependent variable for the excluded category. So, in this specific case, the OHE encoding has a close relationship with the simple target encoder.

Some drawbacks of OHE are widely known. The main problem consists of the tendency of dummy variables to cause overfitting. Moreover, this method produces a lot of new orthogonal dummy variables that slow down and bias the learning if the number of the categories is high, generating a sparse array. Furthermore, the new dummy variables are perfectly independent and this is unrealistic: categories have relationships and similarities that could be extracted from the context and used in the analysis.

Just as an example, consider the "day of the week" variable, which is a cyclic ordinal feature. If the days of the week are represented through 7 one-hot vectors, a spatial representation in 7 dimensions can be obtained, in which the categories are at Euclidean distance $\sqrt{2}$ from each other. This representation is not meaningful: it could be more coherent, for example, to represent the days of the week on a circumference in a smaller two-dimensional space. On the other hand, even a "circular" representation does not take into account that the days of the week can be distinguished between "working" and "non-working", and this distinction is often more relevant for the analysis. In Table 1, we have reported the distances between the days of the week obtained by considering the encodings given by Glove's word-vectors (Pennington et al., 2014) obtained by analyzing millions of documents by a Natural Language Processing model. If our analysis concerns, e.g., the sales forecast of a supermarket or the level of particulates in the air of a great city, this is certainly a coherent coding.

Table 1 Distances between the days of the week in the 50-dimensional Glove word-vectors

Day	Tuesday	Wednesday	Thursday	Friday	Saturday	Sunday
Monday	0.002	0.004	0.002	0.007	**0.086**	**0.074**
Tuesday		0.002	0.002	0.009	**0.087**	**0.077**
Wednesday			0.003	0.008	**0.079**	**0.069**
Thursday				0.007	**0.085**	**0.074**
Friday					**0.072**	**0.062**
Saturday						0.009

Fig. 1 A bidimensional embedding of Italian provinces

As another example, consider to encode the "Italian provinces", that is, a high-cardinality variable. OHE would provide a representation in a big dimensional space while a representation in a two-dimensional space, using, for example, the geographical position of the principal city could be suitable for our analysis (Fig. 1). In general, a high-cardinality variable can have a satisfactory representation in a small space, representing the relationships among the categories, but there is no embedding which is optimal independently of the objective of the analysis and the model applied.

3 Single and Multiple Quantifications by OHE

The encoding of categorical variables has been extensively studied in the approach named Optimal Scaling (OS, Gifi, 1990) where the embedding of the categories in a p-dimensional space was proposed. Given a categorical variable X that can take on the categories $[a_1, a_2, ..., a_k]$, n the number of observations, then $G = [g_1, g_2, ..., g_k]$ is the indicator matrix with dimension $n \times k$. Let $\mathbf{c}$ be a vector of k real values; the generic quantification vector of X is given by

$$\mathbf{x} = \mathbf{Gc} = \sum_{h=1}^{k} c_h \mathbf{g}_h \tag{1}$$

The k values in $\mathbf{c}$ are the quantifications of the k categories and are parameters to be estimated. The quantification vector $\mathbf{x}$ is a linear combination of the indicator variables, which are an orthogonal base of R^k then is defined in a subspace of R^k. The order indicator matrices, with non-negativity constraints on the coefficients, can be used (Gifi, 1990) to obtain ordered quantifications in OS. Considering a supervised approach, if we set the coefficients c_h as the mean of the target Y for the corresponding category, we get the target encoder. In effect, this is the OS solution for only one explanatory unordered categorical variable. With m categorical explanatory variables, MORALS (Young et al., 1976) solves, by an alternating least squares algorithm (ALS, Gifi, 1990), the following expression:

$$\beta_0 + \sum_{j=1}^{m} \beta_j \mathbf{x}_j = \beta_0 + \sum_{j=1}^{m} \beta_j \mathbf{G}_j \mathbf{c}_j =$$

$$= \beta_0 + \sum_{j=1}^{m} \beta_j \sum_{h=1}^{k_j} c_{jh} \mathbf{g}_{jh} = \hat{\mathbf{y}} \tag{2}$$

with the constraints $\mathbf{u}'\mathbf{x}_j = 0$, $\frac{1}{n}\mathbf{x}'_j, \mathbf{x}_j = 1$, $(j = 1, 2, ..., m)$, where $\mathbf{u}$ is the unitary vector.

With only explanatory nominal variables, unless a different normalization of the parameters, MORALS therefore substantially is equivalent to a linear regression with OHE. With few units and high- cardinality variables, this model can lead to even greater overfitting than obtained by the simple target encoder. One possibility to control overfitting in MORALS is the introduction of L1 and L2 penalties on the parameters (Meulman et al., 2019). The strength of this model is that it allows to analyze variables with a mixed measurement level, including also qualitative ordinal variables and transformations of quantitative variables (De Leeuw et al., 1976).

In expression (1), the embedding of one categorical variable in a one-dimensional space R is considered, but we noted in the previous section that a representation in two or more dimensions can be more effective. Considering a regression problem, in

OS (MORALS, 1976), it is possible to obtain multiple quantifications by means of copies of the variables (Gifi, 1990). After choosing the number p of quantifications, we can extend (1) as

$$\underset{n\times p}{\mathbf{X}} = \underset{n\times k}{\mathbf{G}}\ \underset{k\times p}{\mathbf{C}} = \sum_{h=1}^{k} \underset{n\times 1}{\mathbf{g}_h} \cdot \underset{1\times p}{\mathbf{c}_h} \tag{3}$$

with the appropriate constraints. However, this would lead to a worsening of overfitting problems.

A more natural way to consider embedding in two or more dimensions is by Homogeneity Analysis (HOMALS, De Leeuw, 1973). This unsupervised method is equivalent to Multiple Correspondence Analysis (MCA, Benzécri, 1973). With HOMALS or MCA, the categories are "optimally" encoded by maximizing the eigenvalues of the encoded variable correlation matrix. In MCA, the problem is solved analytically, while in HOMALS the problem is solved numerically. This numerical variant offers great flexibility in our case.

Considering m categorical variables, and fixed to p the chosen dimensionality, the HOMALS objective is to find the object scores $\mathbf{Z}$ and quantifications $\mathbf{C}_j$ so that

$$\sigma\left(\mathbf{Z}, \mathbf{C}_1, \mathbf{C}_2, \ldots \mathbf{C}_m\right) = \sum_{j=1}^{m} tr\left(\mathbf{Z} - \mathbf{G}_j\mathbf{C}_j\right)' \left(\mathbf{Z} - \mathbf{G}_j\mathbf{C}_j\right) =$$

$$= \sum_{j=1}^{m} \left\| \mathbf{Z} - \mathbf{G}_j\mathbf{C}_j \right\|^2 = min \tag{4}$$

where $\mathbf{Z}_{(n\times p)}$ is the score matrix, $\mathbf{G}_{j(n\times k_j)}$ is an indicator matrix, and $\mathbf{C}_j$ is the quantification matrix $(k_j \times p)$. The minimization is obtained with the normalization constraints: $\mathbf{u}'\mathbf{Z} = \mathbf{0},\ \mathbf{Z}'\mathbf{Z} = n\mathbf{I}_p$, $\mathbf{I}_p$ is the unitary matrix $(p \times p)$, to avoid the trivial solutions: $\mathbf{Z} = \mathbf{0},\ \mathbf{C}_j = \mathbf{0}$. MCA/HOMALS approaches are linear methods and seek a map in which both units and variables are represented in a low-dimensional Euclidean space in such a way that an observed unit is relatively close to the categories that characterize it and away from the categories that do not characterize it. In this representation, the category embeddings are the centers of gravity of the units that share the same category. In the last section, a non-linear approach is shown, in order to obtain an optimal embedding of the categories.

4 Category Embedding by Neural Networks

Application of Neural Networks to categorical data requires an encoding of the categories. In this paper, an approach to embedding the categories in a low-dimensional space is proposed. First, a supervised approach is considered, in which we have to predict a quantitative target Y. For neural networks, a well-known form of embedding is word-embedding (Bengio et al., 2003). Embedding in NLP is a vector representation of the words in such a way that the words that frequently appear in similar contexts are close to each other. It is common in natural language applications to use 50, 100, or 300 dimensions. However, in our case the vocabulary is composed of the categories of the variable, then the vocabulary is not 50,000 terms but at most tens or hundreds so a dimension of 2 or 3 can often be sufficient. Each category is mapped to a distinct vector, which is learned while training the neural network.

In our approach, quantification (1) is defined as a multi-input neural network, where each categorical variable is a distinct OHE input and is followed by one dense layer with p neurons (the chosen dimension) without bias and with linear activation function. In the successive layer, the outputs, coming from all the variables, must be concatenated. Inspired by Natural Language Processing, Guo and Berkhahn's entity embedding technique Guo and Berkhahn (2016) considered a similar approach.

For example, consider 3 input categorical variables, each with 100 categories, and one hidden layer containing 512 neurons; using this approach, we must estimate (considering a regression problem and $p = 2$) 4.697 weights. In fact, given $t = 512$, $p = 2, m = 3, k_j = 100$ for each j, the Neural Network can be written as

$$\hat{\mathbf{y}} = \beta_0 + \sum_{s=1}^{t} \beta_s \phi \left(\sum_{j=1}^{m} \sum_{r=1}^{p} \mathbf{G}_j \mathbf{c}_j^r w_{jrs} + w_{0s} \right) \tag{5}$$

where ϕ is the activation function of the hidden layer, and $\mathbf{c}_j^r$ is the quantification of the $j-th$ variable on the $r-th$ dimension. Conversely, in the classical OHE encoding:

$$\hat{\mathbf{y}} = \beta_0 + \sum_{s=1}^{t} \beta_s \phi \left(\sum_{j=1}^{m} \sum_{r=1}^{k_j} \mathbf{G}_j w_{jrs} + w_{0s} \right) \tag{6}$$

obtaining 154.625 weights to estimate.

$\mathbf{G}_j$ can be a very big sparse matrix (sparsity equal to $1 - 1 - -k_j$), but it is possible to build the dense matrix of quantifications $\mathbf{c}_j^r$ of expression (5) without building $\mathbf{G}_j$. In the first step, for each categorical variable X, the k-dimensional "vocabulary" V of the categories have to be created and indexed. Then all the categories in the data will be substituted by the corresponding numerical index in the vocabulary, in a similar way to what the Label Encoder does. Call a_i the category assumed by the categorical variable on the i-th unit, and $\mathbf{v}[a_i]$ the index in the vocabulary corresponding to this category; the i-th row of the quantified variable X can be expressed as

$$\mathbf{x}_i = \mathbf{C}[\mathbf{v}[a_i]] \tag{7}$$

To obtain the estimate of $\mathbf{C}$ in a supervised neural network, the gradient descent and the backpropagation can be used, where the matrix $\mathbf{C}$ is initialized with random values taken from a standardized normal and subsequently updated through an iterative procedure to minimize the loss function, which in the case of regression is the classic

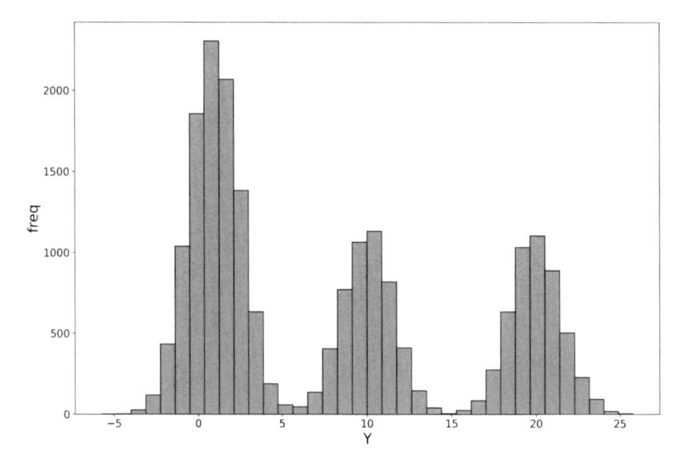

Fig. 2 Distribution of the target Y in the simulation

Sum of Square Error. We call this technique LEE, Low Embedding Encoder, and a small simulation for a regression problem was built to illustrate the proposed approach. Consider three categorical variables X_1, X_2, X_3 with 200 categories each (indicated by the integers between 1 and 200); 20,000 observations were extracted randomly for each variable from a uniform distribution, then Y was computed by the rules:

$$\begin{array}{l} if\ X_1 > X_2\ and\ X_3 < 100 \Rightarrow Y \sim N(20, 1.5) \\ if\ X_1 \leq X_2\ and\ X_3 < 100 \Rightarrow Y \sim N(10, 1.5) \\ in\ all\ other\ cases\ \ Y \sim N(1, 1.5) \end{array}$$

There are only 3 expected values $E(Y|x_1, x_2, x_3)$, i.e., (1, 10, 20), so an optimal regression model should predict these values. Note that the expected value of Y depends on the interaction of the three categorical variables and that the three conditional distributions of Y overlap in the tails (Fig. 2). The dataset was then split as training set (50%) and test set (50%). Regression algorithms such as MORALS or Regression Tree cannot make a satisfactory prediction on this data unless introducing explicitly all the interaction terms into the model, producing thousands of dummy variables. On the contrary, neural networks can autonomously detect the interactions among the variables. Then, for this simulation, after encoding the categorical variables, a small neural network was chosen to predict the target Y. The neural network includes an input layer, two hidden layers with 8 and 3 neurons (ELU activation function), and 1 output neuron with linear activation function. As for the data leakage, which is always possible in the supervised case, this can be controlled with the usual tools of neural networks, as simplify the architecture introduces dropout layers or L1/L2 penalizations.

Applying the LEE approach, each categorical variable is considered a separate input coded as OHE, followed by one dense layer with 2 neurons ($p = 2$) and no bias. If we want to avoid the sparse coding matrices of OHE, an embedding layer can

Table 2 Comparison between three encoding approaches (200 iterations)

	MSE-train	MSE-test	n.parameters
OHE	2.11	6.18	4839
LEE	2.62	4.64	1287
Target encoder	7.22	7.29	1673

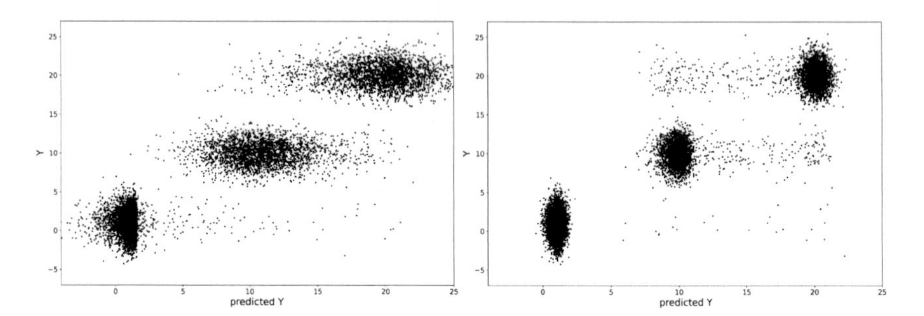

Fig. 3 Distribution of the target Y in the simulation for OHE (left) and LEE (right)

be used, for each original categorical variable, without creating the dummy variables (for example, the embedding layer of the software library TensorFlow is available https://www.tensorflow.org/). To verify that the results obtained, shown in Table 2, do not depend on the choice of the supervised model applied, a check was also made with other network architectures. The Target Encoder obtained bad results also on the training set but with far fewer parameters (only 63), then this encoder was applied with a bigger neural network, with 38 neurons in each hidden layer, to obtain an equivalent number of parameters in the comparison. Despite the increase in the number of parameters, the result is poor even on the training set, since this encoding does not allow the identification of interactions.

In Fig. 3, it is possible to compare the performance of OHE and LEE encoders in allowing the neural network to identify the expected values of Y. In particular, the true Y values (ordinates) and the estimated Y values (abscissas) are shown, for the OHE (left) and LEE (right). In the optimal result, the predicted Y should have only 3 values: 1, 10, and 20.

5 Non-linear Encoding in the Unsupervised Case

Using an autoencoder neural network, it is also possible to reproduce, with some efforts, the quantifications provided by HOMALS. To obtain this result, it is necessary to build an encoder-average-decoder NN with three internal layers to reproduce the loss function (4). The encoder will consist of the LEE encoding, i.e., the introduction of variables through OHE, then a small dense layer with linear activation for each

Table 3 A frequency table showing association between X and Y

X/Y	a	b	c	d	e	Total
A	801	100	100	100	100	1201
B	100	800	100	100	100	1200
C	100	100	800	100	100	1200
D	100	100	100	800	100	1200
E	100	100	100	100	800	1200
Total	1201	1200	1200	1200	1200	6001

variable with size = p, then one average layer. The decoder will be constituted inversely to the encoder, with, in the output, the variables encoded as OHE and the same parameters as the encoder. Finally, to obtain the same result of (4), it would be also necessary to impose orthogonality and normalization constraints followed by an appropriate rotation of the components obtained. Obviously, all this would not make sense, as with much less effort we can use the elegant analytical solution provided by MCA.

On the other hand, this scheme can be easily modified to obtain a non-linear solution simply by using non-linear activation functions in the decoder instead of linear functions, without imposing any constraints on the parameters. The coding achieved in this way can be much more effective than that provided by HOMALS/MCA.

As an example, let's consider only two categorical variables, X and Y, each with 5 categories, respectively (A, B, C, D, E) and (a, b, c, d, e), which give rise to the frequency Table 3. The strong associations of the pairs of categories (A, a), (B, b), (C, c), (D, d), (E, e) are evident from the table. We would therefore expect a representation on two components that highlights these associations, which is the main information inside the table: a representation in which those pairs are close while there is an approximate equidistance from the other categories. By applying the MCA, the first 4 components have the same eigenvalue and are all necessary to have a satisfactory representation of the categories. Figure 4 shows the result obtained on the first 2 components with MCA (on the left) and with the non-linear version just described (on the right). Note how the presence of only one more unit for the pair (A, a) completely distorts the result of MCA.

6 Conclusions

The proposed encoding method LEE allows to apply neural networks to data with categorical variables and high cardinality, reducing the number of parameters and memory resources. The simulation results obtained show an increased predictive

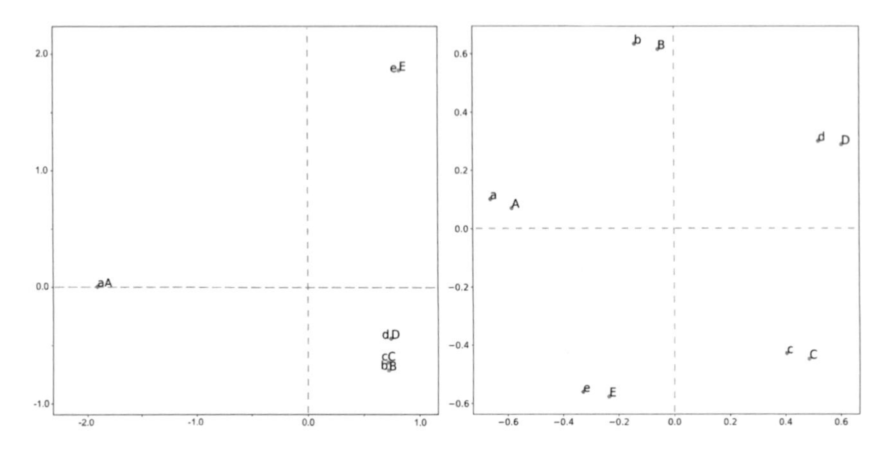

Fig. 4 Categorical encoding for Correspondence Analysis (left) and LEE (right) on data of Table 3

capacity of the neural network, thanks to the more efficient encoding. Even in the unsupervised case, this approach could provide more effective and interpretable non-linear quantifications than classical methods.

References

Bengio, Y., Ducharme, R., Vincent, P., & Janvin, C. (2003). A neural probabilistic language model. *The Journal of Machine Learning Research, 3*, 1137–1155.

Benzécri, J. P. (1973). *L'Analyse Des Données: Tome II: L'Analyse Des Correspondances*. Paris: Dunod.

De Leeuw, J. (1973). Canonical Analysis of Categorical Data. Ph.D. thesis, University of Leiden, The Netherlands (1973).

De Leeuw, J., Young, F. W., & Takane, Y. (1976). Additive structure in qualitative data: an alternating least squares method with optimal scaling features. *Psychometrika, 41*(4), 471–503.

Di Ciaccio, A. (2020). Categorical encoding for machine learning. In A. Pollice et al. (Eds.), *Book of short papers SIS2020*. Pearson Italia.

Gifi, A. (1990). *Nonlinear Multivariate Analysis*. New York: Wiley.

Guo, C., & Berkhahn, F. (2016). *Entity embeddings of categorical variables*. arXiv:1604.06737.

Hancock, J. T., & Khoshgoftaar, T. M. (2020). Survey on categorical data for neural networks. *Journal of Big Data, 7*, 28. https://doi.org/10.1186/s40537-020-00305-w.

https://scikit-learn.org/ .

https://www.tensorflow.org/ .

Meulman, J., van der Kooij, A. J., & Duisters, K. L. W. (2019). ROS Regression: integrating regularization with optimal scaling regression. *Statistical Science, 34*(3), 361–390.

Pennington, J., Socher, R., & Manning, C. D. (2014). GloVe: global vectors for word representation. *Proceedings of the Conference on Empirical Methods in Natural Language Processing (EMNLP)*. https://doi.org/10.3115/v1/D14-1162

Potdar, K., Pardawala, T. S., & Pai, C. D. (2017). A comparative study of categorical variable encoding techniques for neural network classifiers. *International Journal of Computer Applications, 175*, 4.

Yong-Xia, Z., & Ge, Z. (2010). MD5 research. In *2010 Second International Conference on Multimedia and Information Technology*, Kaifeng (pp. 271–273).
Young, F. W., De Leeuw, J., Takane, Y. (1976). Regression with qualitative and quantitative variables: An alternating least squares method with optimal scaling features. *Psychometrika, 41*(4).

Bayesian Multivariate Analysis of Mixed Data

Chiara Galimberti, Federico Castelletti, and Stefano Peluso

Abstract Graphical models provide an effective tool to represent conditional independences among variables. While this class of models has been extensively studied in the Gaussian and categorical settings separately, state-of-the-art literature which jointly models the two types of data is narrow. However, mixed data are widespread in many applications where both continuous and categorical measurements are available, such as genomics or industrial machine processes. In this paper, we propose a Bayesian framework for the analysis of mixed data. Specifically, we define a likelihood function for n observations following a conditional Gaussian distribution and assign suitable priors to model parameters. We develop an MCMC scheme for approximate posterior inference in two alternative model parameterizations and a structure learning algorithm for related undirected graph models.

Keywords Conditional Gaussian distribution · Undirected graph · Graphical model · Marginal likelihood · Mixed variables · Structure learning

C. Galimberti (✉)
Department of Economics, Management and Statistics, Universitá degli Studi di Milano-Bicocca, Milan, Italy
e-mail: c.galimberti19@campus.unimib.it

F. Castelletti
Department of Statistical Sciences, UniversitÃ Cattolica del Sacro Cuore, Milan, Italy
e-mail: federico.castelletti@unicatt.it

S. Peluso
Department of Statistics and Quantitative Methods, Universitá degli Studi di Milano-Bicocca, Milan, Italy
e-mail: stefano.peluso@unimib.it

L. Grilli et al. (eds.), *Statistical Models and Methods for Data Science*, Studies in Classification, Data Analysis, and Knowledge Organization,
https://doi.org/10.1007/978-3-031-30164-3_5

1 Introduction

Graphical models provide a powerful framework to represent conditional dependence structures in multivariate distributions (Lauritzen, 1996). Typically the graphical model generating the observations is unknown and inferring it from the data is possible using *structure learning* methodologies. Several contributions for structure learning of graphical models given continuous (Gaussian) or discrete/categorical data are available in the literature; see for instance Kalisch and Bühlmann (2007) and Meinshausen and Bühlmann (2006). However, mixed-type data are extremely diffuse in many contexts where both continuous and categorical measurements are available. One instance is nanotechnology, which develops functional structures designed at the atomic or molecular scale, related to optoelectronics, luminescent materials, lasing materials, and biomedical imaging Hodes et al. (1987), Bawendi et al. (1989), and Ma et al. (2004). In particular, nanostructure data may be represented as categorical data, while the variables involved in the process of measuring the nanostructures are continuous.

However, literature oriented to graphical models for mixed-type data is narrow, especially in the Bayesian framework. A few recent works on graphical models for mixed data with a frequentist approach are available in the literature. Some of them propose parameter estimation methods for models belonging to exponential families based on node-wise conditional generalized linear regressions; see Chen et al. (2015), Yang et al. (2012), and Yang et al. (2014). Other contributions work in the context of the Conditional Gaussian (CG) distribution introduced by Lauritzen and Wermuth (1989); in particular, Lee and Hastie (2015) propose a maximum pseudo-likelihood approach under the assumption of constant conditional covariance for continuous variables, while Cheng et al. (2017) fit separate regression models for each variable with a weighted lasso penalty. The more recent Bayesian methodology of Bhadra et al. (2018) allows to infer a network structure given mixed data by adopting Gaussian scale mixtures, while Zareifard et al. (2021) propose a Gibbs sampler algorithm to estimate a Directed Acyclic Graph (DAG) in the presence of continuous and discrete observations.

In this contribution, we focus on undirected graphs (UGs). The scope of our study is to develop a Bayesian methodology for structure learning and inference of UG models from mixed data. The starting point of our model is the CG distribution. We elaborate on the latter model by extending it to a Bayesian framework and providing a class of prior distributions for the model parameters. Specifically, we adopt two different parameterizations for the same model: a *moment* and a *canonical* representation; the first one enables closed-form results in terms of parameter posterior distribution as well as marginal likelihood. The second instead provides an effective way to express conditional independence relations in the joint distribution, and in turn to learn the underlying graphical structure; from a computational perspective, it requires Markov Chain Monte Carlo (MCMC) strategies for approximate posterior inference.

The rest of the contribution is organized as follows. In Sect. 2.1, we present some results obtained through the moment representation of the model, while in Sect. 2.2 we introduce the canonical parameterization. In Sect. 3, we present an application to heart disease diagnosis data, and we conclude in Sect. 4 with a discussion.

2 Bayesian Model Development

The starting point of our model is represented by the definition of Conditional Gaussian (CG) distribution, as introduced by Lauritzen and Wermuth (1989). Let V be a finite set of nodes indexing a collection of random variables $X = (X_1, \ldots, X_{|V|})^T$, which includes both discrete and continuous quantities indexed by $\Delta \cup \Gamma = V$, respectively. Lauritzen and Wermuth (1989) defined a general class of probability distributions of the form

$$f(\mathbf{x}) = f(s, \mathbf{y}) = \exp\left\{g(s) + \mathbf{h}(s)^T\mathbf{y} - \frac{1}{2}\mathbf{y}^T\mathbf{K}(s)\mathbf{y}\right\} \tag{1}$$

where s and y correspond to the multi-dimensional levels assumed by the categorical and continuous variables, respectively. A probability distribution of the form (1) has CG distribution if and only if $X_\Gamma | X_\Delta = s \sim \mathcal{N}_{|\Gamma|}(\mathbf{K}(s)^{-1}\mathbf{h}(s), \mathbf{K}(s)^{-1})$ and the marginal distribution of the discrete variables is

$$\theta(s) = (2\pi)^{-\frac{|\Gamma|}{2}} |\mathbf{K}(s)|^{-\frac{1}{2}} \exp\left\{g(s) + \frac{1}{2}\mathbf{h}(s)^T\mathbf{K}(s)^{-1}\mathbf{h}(s)\right\}, \tag{2}$$

for each level s assumed by Z_Δ. Moreover, if $\mathbf{K}(s) = \mathbf{K}$ a CG distribution is called *homogeneous* (HCG). A representation based on the triplet $(g, \mathbf{h}, \mathbf{K})$ is named *canonical*. One possible alternative parameterization is given in terms of *moment-characteristics* parameters $(\boldsymbol{\theta}, \boldsymbol{\mu}, \boldsymbol{\Omega})$.

In the following sections, we detail some results for a Bayesian model formulation under both parameterizations.

2.1 *Moment Representation*

Let $(Z_1, \ldots, Z_p)$ be p categorical variables, $(Y_1, \ldots, Y_q)$ q continuous variables. Let also $\mathcal{I}$ be the space of all possible configurations of the p categorical variables and $\boldsymbol{\theta} = \{\theta(s), s \in \mathcal{I}\})$ where $\theta(s) = \Pr(Z_1 = s_1, \ldots Z_p = s_p)$ is the probability to observe configuration $s = (s_1, \ldots, s_p)$.

Under the HCG assumption, we can write for each $s \in \mathcal{I}$

$$Y_1(s), \ldots, Y_q(s) \mid \boldsymbol{\mu}(s), \boldsymbol{\Omega} \sim \mathcal{N}_q(\boldsymbol{\mu}(s), \boldsymbol{\Omega}^{-1}). \tag{3}$$

In particular, the relation between the canonical representation and the moments of the Gaussian model can be expressed through the re-parameterization $\boldsymbol{\mu}(s) = \mathbf{K}^{-1}\mathbf{h}(s)$ and $\boldsymbol{\Omega} = \mathbf{K}^{-1}$.

Consider now a collection of n i.i.d. observations $z_i = (z_{i,1}, \ldots, z_{i,p})^T$, $\boldsymbol{y}_i = (y_{i,1}, \ldots, y_{i,q})^T$, $i = 1, \ldots, n$ from (2) and (3). Categorical data $\{\boldsymbol{x}_i, i = 1, \ldots, n\}$ can be equivalently represented as a contingency table of counts $\boldsymbol{N}$ with elements $n(s) \in \boldsymbol{N}$ such that

$$n(s) = \sum_{i=1}^{n} \mathbf{1}(z_i = s),$$

where $\mathbf{1}(\cdot)$ is the indicator function and $\sum_{s \in \mathcal{I}} n(s) = n$. Following Frydenberg & Lauritzen (1989), the likelihood function can be written as

$$\begin{aligned} f(\boldsymbol{N}, \boldsymbol{y}_1, \ldots, \boldsymbol{y}_n \mid \boldsymbol{\theta}, \{\boldsymbol{\mu}(s)\}_{s \in \mathcal{I}}, \boldsymbol{\Omega}) &= \prod_{s \in \mathcal{I}} \theta(s)^{n(s)} \prod_{s \in \mathcal{I}} \prod_{i \in \nu(s)} \phi(\boldsymbol{y}_i \mid \boldsymbol{\mu}(s), \boldsymbol{\Omega}^{-1}) \\ &\propto \prod_{s \in \mathcal{I}} \theta(s)^{n(s)} \prod_{s \in \mathcal{I}} \prod_{i \in \nu(s)} |\boldsymbol{\Omega}|^{\frac{1}{2}} \exp\left\{ -\frac{1}{2} (\boldsymbol{y}_i - \boldsymbol{\mu}(s))^T \boldsymbol{\Omega} (\boldsymbol{y}_i - \boldsymbol{\mu}(s)) \right\}, \end{aligned} \tag{4}$$

where $\nu(s)$ is the set of observations among $\{1, \ldots, n\}$ with observed configuration s and ϕ denotes the Gaussian density. We complete our Bayesian model formulation by assigning the following prior distributions:

$$\boldsymbol{\theta} \sim \text{Dirichlet}(\boldsymbol{A}), \quad \boldsymbol{\mu}(s) \mid \boldsymbol{\Omega} \sim \mathcal{N}_q(\boldsymbol{m}(s), (a_\mu \boldsymbol{\Omega})^{-1}), \quad \boldsymbol{\Omega} \sim \mathcal{W}_q(a_\Omega, \boldsymbol{U}), \tag{5}$$

where in particular $\mathcal{W}_q(a_\Omega, \boldsymbol{U})$ denotes a Wishart distribution having expectation $a_\Omega \boldsymbol{U}^{-1}$, with $a_\Omega > q - 1$ and $\boldsymbol{U}$ a s.p.d. matrix.

It is advisable to set hyperparameters to values leading to proper prior distributions. A standard way to proceed, whenever no substantial prior information is available, is to choose hyperparameters leading to weakly informative priors. In particular, $\boldsymbol{A}$ may be set equal to a vector with all equal (e.g. unit) components (each associated to one level of the categorical variables). With regard to the Normal priors, we can fix a zero mean, while $a_\mu = 1$. Finally, the hyperparameters of the Wishart distribution can be fixed as $a_\Omega = q$, $\boldsymbol{U} = \boldsymbol{I}_q$, the (q, q) the identity matrix.

Under prior parameter independence, the posterior distribution can be written after standard calculations as

$$p(\boldsymbol{\theta}, \{\boldsymbol{\mu}(s)\}_{s\in\mathcal{I}}, \boldsymbol{\Omega} \mid \boldsymbol{N}, \boldsymbol{y}_1, \ldots, \boldsymbol{y}_n) \propto \prod_{s\in\mathcal{I}} \theta(s)^{a(s)+n(s)-1}$$

$$\cdot \prod_{s\in\mathcal{I}} \left\{ |\boldsymbol{\Omega}|^{\frac{1}{2}} | \exp\left\{ -\frac{1}{2}(n(s)+a_\mu)(\boldsymbol{\mu}(s)-\bar{\boldsymbol{m}}(s))^T \boldsymbol{\Omega}(\boldsymbol{\mu}(s)-\bar{\boldsymbol{m}}(s)) \right\} \right.$$

$$\cdot |\boldsymbol{\Omega}|^{\frac{a_\Omega+n-q-1}{2}} \exp\left\{ -\frac{1}{2}\mathrm{tr}[(\boldsymbol{U}+\boldsymbol{R}+\boldsymbol{R}_0)\boldsymbol{\Omega}] \right\}, \tag{6}$$

with $\boldsymbol{R} = \sum_{s\in\mathcal{I}} \mathrm{SSD}(s)$,

$$\bar{\boldsymbol{m}}(s) = \frac{a_\mu}{a_\mu + n(s)}\boldsymbol{m}(s) + \frac{n(s)}{a_\mu + n(s)}\bar{\boldsymbol{y}}(s),$$

$$\boldsymbol{R}_0 = \sum_{s\in\mathcal{I}} \frac{a_\mu n(s)}{a_\mu + n(s)}(\boldsymbol{m}(s) - \bar{\boldsymbol{y}}(s))(\boldsymbol{m}(s) - \bar{\boldsymbol{y}}(s))^T,$$

where $\mathrm{SSD}(s) = \sum_{i\in\nu(s)} \boldsymbol{e}_i\boldsymbol{e}_i^T$, $\boldsymbol{e}_i = (\boldsymbol{y}_i - \bar{\boldsymbol{y}}(s))$, and $\bar{\boldsymbol{y}}(s)$ is the $(q, 1)$ vector with sample means of $(Y_1, \ldots, Y_q)$ relative to observations $i \in \nu(s)$. Note that the independence is assumed also among the configuration s for the whole set of $\boldsymbol{\mu}$ in order to have prior independence between $\boldsymbol{\mu}(s)$'s. From Eq. (6), it follows that

$$\begin{aligned} \boldsymbol{\theta} \mid \boldsymbol{N} &\sim \mathrm{Dirichlet}(\boldsymbol{A} + \boldsymbol{N}) \\ \boldsymbol{\mu}(s) \mid \boldsymbol{N}, \boldsymbol{Y}, \boldsymbol{\Omega} &\sim \boldsymbol{N}_q(\bar{\boldsymbol{m}}(s), [(a_\mu + n(s))\boldsymbol{\Omega})]^{-1}) \\ \boldsymbol{\Omega} \mid \boldsymbol{Y} &\sim \mathcal{W}_q(a_\Omega + n, \boldsymbol{U} + \boldsymbol{R} + \boldsymbol{R}_0), \end{aligned} \tag{7}$$

where $\boldsymbol{Y}$ denotes the (n, q) data matrix, row-binding of the $\boldsymbol{y}_i$'s; see also Degroot (2004) for details on multivariate Normal models with Normal-Wishart priors and posterior calculations.

In addition, because of conjugacy, the marginal data distribution

$$m(\boldsymbol{Y}, \boldsymbol{N}) = \int f(\boldsymbol{N}, \boldsymbol{Y} \mid \boldsymbol{\theta}, \{\boldsymbol{\mu}(s)\}_{s\in\mathcal{I}}, \boldsymbol{\Omega}) p(\boldsymbol{\theta}) \prod_{s\in\mathcal{I}} p(\boldsymbol{\mu}(s)) p(\boldsymbol{\Omega})\, d\boldsymbol{\theta} \prod_{s\in\mathcal{I}} d\boldsymbol{\mu}(s)\, d\boldsymbol{\Omega}$$

can be computed from the ratio of prior and posterior normalizing constants.

The result in (7) enables posterior inference on the parameters of an *unconstrained* (complete) graphical model. Specifically, by implementing a Monte Carlo sampler it is possible to infer the parameters of the marginal distribution of discrete variables and the parameters of the conditional distribution of continuous (Gaussian) variables.

2.2 Canonical Representation

The canonical parameterization of a CG distribution relies on the notion of *interaction*. In particular, the triplet $(g, \mathbf{h}, \mathbf{K})$ in (1) can be first expressed through the following expansions:

$$g(s) = \sum_{d:d\subseteq\Delta} \lambda_d(s), \qquad \mathbf{h}(s) = \sum_{d:d\subseteq\Delta} \boldsymbol{\eta}_d(s), \qquad \mathbf{K}(s) = \sum_{d:d\subseteq\Delta} \boldsymbol{\Phi}_d(s), \tag{8}$$

where parameters λ, $\boldsymbol{\eta}$, and $\boldsymbol{\Phi}$ are called *interaction* terms and d represents any subset (including the null set) of the categorical variables. Each term of the expansions represents a type of interaction between variables. Specifically,

- $\lambda_\emptyset$ is the *log normalizing constant*. Also, λ_d $(d \neq \emptyset)$ are *pure discrete interactions* among variables in d. If $|d| = 1$ they correspond to the *main effects* of the discrete variables;
- $\boldsymbol{\eta}_\emptyset$'s coordinates are the *main effects* of the continuous variables. Instead, $\boldsymbol{\eta}_d$, $d \neq \emptyset$ are *mixed linear interactions* between a continuous variable and variables in d;
- $\boldsymbol{\Phi}_\emptyset$'s elements are *pure quadratic interactions*. Differently, $\boldsymbol{\Phi}_d$ $(d \neq \emptyset)$ are *mixed quadratic interaction matrices* between variables in d and pairs of continuous variables.

Using this representation, the distribution is homogeneous if and only if it has an interaction representation with no mixed quadratic interactions. Interaction terms allow for a more direct characterization of conditional independencies between variables; accordingly, the *Markov property* of a given UG can be expressed through zero-constraints on such parameters. Let $\mathcal{G}$ be an UG; a CG distribution is said to be *nearest-neighbour Gibbs* with respect to a graph $\mathcal{G}$ (or $\mathcal{G}$-Gibbsian) if it has a representation with interaction terms satisfying

$$\begin{aligned} \lambda_d(s) &\equiv 0 \quad \text{unless } d \text{ is complete in } \mathcal{G}, \\ \eta_d(s)_\gamma &\equiv 0 \quad \text{unless } d \cup \{\gamma\} \text{ is complete in } \mathcal{G}, \\ \boldsymbol{\Phi}_d^{\gamma\delta}(s) &\equiv 0 \quad \text{unless } d \cup \{\gamma, \delta\} \text{ is complete in } \mathcal{G}, \end{aligned} \tag{9}$$

where γ and δ represent continuous variables in Γ.

Notice that a Gibbsian probability has an expansion with interaction terms involving variables that are neighbours only. Moreover, it can be proved that a CG distribution is $\mathcal{G}$-Markovian if and only if it is $\mathcal{G}$-Gibbsian; see Proposition 3.1 in Lauritzen and Wermuth (1989). As a consequence, the joint density factorizes into a product of local densities that only depends on variables that are mutual neighbours. In addition, it can be shown that the so-obtained factorization splits up into separate factorizations of the constant, linear, and quadratic terms; see Appendix B in Lauritzen and Wermuth (1989).

In what follows, we consider a simplified model by imposing the following conditions on the order of interactions:

- $|d| \leq 2$ for $\lambda_d(s)$,
- $|d| \leq 1$ for $\boldsymbol{\eta}_d(s)$,
- $d = \emptyset$ for $\boldsymbol{\Phi}_d(s) \equiv \boldsymbol{\Phi}_d$.

In other words, the simplified model omits all interaction terms between the categorical variables of order higher than two and it defines the canonical mean vector of the Gaussian variables as a linear function of the categorical variables instead of an "arbitrary" dependence function. Moreover, the so-obtained distribution is HCG since the conditional covariance matrix does not depend on the categorical variables. As a consequence, the conditional independencies can be read-off according to the following equivalences:

$$\begin{aligned} Z_j \perp Z_k | X \setminus \{Z_j, Z_k\} &\Leftrightarrow \lambda_{jk} \equiv 0, \\ Z_j \perp Y_\gamma | X \setminus \{Z_j, Y_\gamma\} &\Leftrightarrow \boldsymbol{\eta}_{j\gamma} \equiv 0, \\ Y_\gamma \perp Y_\delta | X \setminus \{Y_\gamma, Y_\delta\} &\Leftrightarrow \boldsymbol{\Phi}_0^{\gamma\delta} \equiv 0, \end{aligned} \tag{10}$$

where (Z_j, Z_k) represents two discrete variables and (Y_γ, Y_δ) represents two continuous variables.

As in the moment representation setting, we consider a mixed dataset consisting of n i.i.d. observations and represent categorical data through the implied contingency table of counts. For simplicity of exposition, in the following, we assume all categorical variables being binary.

By adopting the canonical parameterization, the likelihood function can be written as

$$f(\mathbf{x}_1, \ldots, \mathbf{x}_n | \boldsymbol{\theta}) = \prod_{i=1}^{n} f(\mathbf{x}_i | \boldsymbol{\theta}) \tag{11}$$

$$= \prod_{i=1}^{n} \exp\left\{ [\lambda_0 + \sum_{j=1}^{p} \lambda_j z_{ij} + \sum_{j<k} \lambda_{jk} z_{ij} z_{ik}] + \mathbf{y}_i^T [\eta_0 + \sum_{j=1}^{p} \eta_j z_{ij}] - \frac{1}{2} \mathbf{y}_i^T \boldsymbol{\Phi}_0 \mathbf{y}_i \right\}$$

$$= \exp\left\{ n\lambda_0 + \sum_{j=1}^{p} \lambda_j n(j) + \sum_{k<j} \lambda_{kj} n(k, j) + n\boldsymbol{\eta}_0{}^T \bar{\mathbf{y}} + \sum_{j=1}^{p} \boldsymbol{\eta}_j{}^T \mathbf{t}(j) - \frac{1}{2} \mathrm{tr}(\boldsymbol{R}\boldsymbol{\Phi}_0) \right\},$$

where $n(j)$ and $n(j, k)$ are marginal and joint frequencies corresponding to configurations of the categorical variables, $\bar{\mathbf{y}}$ is the vector-sample mean of the continuous variables, $\mathbf{t}(j)$ represents the interaction vector between the two types of variables, and $\boldsymbol{R}$ is the covariance matrix of continuous variables.

Based on this model specification, it is possible to obtain a closed-form expression for the log normalizing constant λ_0, simply by integrating over $\mathbf{y}$ and summing over the domain $\mathcal{Z}$ of the categorical variables. In the case of HCG distribution, we obtain

$$\exp(\lambda_0)^{-1} = \sum_{\mathbf{z}\in\mathcal{Z}} P(\mathbf{Z}=\mathbf{z})$$
$$= (2\pi)^{-\frac{q}{2}} \det(\mathbf{K})^{-\frac{1}{2}} \sum_{\mathbf{z}\in\mathcal{Z}} \exp\left(g(\mathbf{z}) + \frac{1}{2}\mathbf{h}(\mathbf{z})^T\mathbf{K}^{-1}\mathbf{h}(\mathbf{z})\right), \quad (12)$$

where $P(\mathbf{Z}=\mathbf{z})$ is the marginal distribution of $\mathbf{Z}$ obtained after integration over $\mathbf{y}$. Therefore, the value of λ_0 can be obtained as

$$\lambda_0 = -\log\big(\sum_{\mathbf{z}\in\mathcal{Z}} P(\mathbf{Z}=\mathbf{z})\big). \quad (13)$$

Given the canonical parameterization, we need to specify a set of prior distributions for all the interaction terms. Specifically, we assign to λ_j, $\lambda_{j,k}$, $\boldsymbol{\eta}_0$, and $\boldsymbol{\eta}_j$ for $j, k = 1, \ldots, p$ $(j < k)$ zero-mean multivariate Normal distributions, while a Wishart prior to $\boldsymbol{\Phi}_0$.

This type of representation allows for posterior inference on model parameters and implicitly performs structure learning of the underlying UG according to (9). Through the implementation of a suitable MCMC scheme, it is possible to approximate the posterior distribution of each interaction term parameter, whose posterior mean provides a point estimate of the corresponding term. Moreover, structure learning can be performed by building a credible set on the posterior distribution of each interaction term. For a one-dimensional parameter β, a credible interval of level $(1-\alpha)$ is the symmetric interval $I_\alpha(\beta)$ which contains a proportion $(1-\alpha)$ of the probability mass of the posterior distribution. Given that the presence of an edge corresponds to a non-zero value of the associated parameter, the adjacency matrix $\boldsymbol{G}$ of graph $\mathcal{G}$ can be estimated by excluding edges whenever the associated credible interval includes the zero value. More in detail, starting from the MCMC output, we recover $\hat{I}_\alpha(\lambda_{jk})$, $\hat{I}_\alpha(\boldsymbol{\Phi}_{0,jk})$, and $\hat{I}_\alpha(\boldsymbol{\eta}_{jk})$ using the corresponding empirical quantiles. Then $\widehat{\boldsymbol{G}}_{jk} = 1 - \mathbf{1}\left(\hat{I}_\alpha(\lambda_{jk}) \ni 0\right)$ for those edges expressing dependencies between discrete variable, $\widehat{\boldsymbol{G}}_{jk} = 1 - \mathbf{1}\left(\hat{I}_\alpha(\boldsymbol{\Phi}_{0,jk}) \ni 0\right)$ for edges between continuous variables, and $\widehat{\boldsymbol{G}}_{jk} = 1 - \mathbf{1}\left(\hat{I}_\alpha(\boldsymbol{\eta}_{jk}) \ni 0\right)$ for mixed dependencies; compare also with Eq. (10).

3 Real Data Application

In this section, we apply the proposed methodology to a dataset relative to heart disease patients collected from the *Cleveland Clinic Foundation* and publicly available at https://archive.ics.uci.edu/. The original dataset contains $n = 303$ observations and 76 attributes. We include in our study 12 variables, which also received particular attention in several previous analyses. To ease the implementation of our methodology, categorical variables with a number of levels larger than two were also converted into binary variables. Each of the so-obtained categorical attributes thus indicates the absence/presence of a symptom (e.g. "chest pain type"), patient's characteristic (e.g. "fasting blood sugar"), or disease (e.g. "diagnosis of heart disease"). Variables included in the analysis are then divided into the following:

- Gaussian: *age*, resting blood pressure (*trestbps*), serum cholesterol (*chol*), and maximum heart rate achieved (*thalach*);
- Binary: *sex*, fasting blood sugar (*fbs*), exercise induced angina (*exang*), chest pain type (*cp*), diagnosis of heart disease (*num*), resting electrocardiograph results (*restecg*), slope of the peak exercise ST segment (*slope*), and type of test (*thal*).

We approximate the posterior distribution of the model parameters by resorting to the MCMC strategy provided by the R package `MCMCPack` (Martin et al. 2011) which is based on a random walk Metropolis-Hastings (MH) algorithm. Since we are interested in estimating the graphical structure underlying the data, we consider the canonical model representation. Accordingly, starting from the likelihood function in (11), we assign independent Gaussian prior distributions to all the parameters λ and η, and a Wishart prior to the matrix $\boldsymbol{\Phi}_0$. In line with Sect. 2, we fix zero mean and standard deviation equal to one in the Gaussian priors, while the Wishart prior hyperparameters are set equal to $a_\Omega = q$ and $\boldsymbol{U} = \boldsymbol{I}_q$.

The Metropolis-Hastings (MH) algorithm applies to parameter posterior distributions from which direct sampling is not possible. Accordingly, the output is a sequence of random samples approximately drawn from the posterior distribution. At each step, the algorithm proceeds by drawing candidate values from a proposal distribution (Gaussian in our case), and the proposed parameter is accepted with a probability given by the MH ratio.

To initialize the algorithm, we first optimize the posterior distribution with the BFGS (quasi-Netwon) method. We input the so-obtained optimal values as the initial point for the Metropolis-Hastings and adopt the Hessian matrix of the loss function as the covariance matrix of the proposal distribution. We run the MCMC algorithm for 50000 iterations, with a burn-in period of 10000 and a thinning parameter equal to 10. The burn-in period allows to discard some initial samples when the chains are not stationary, and with a thinning parameter equal to 10 we keep one draw every ten values, discarding all other values and avoiding autocorrelations in the chains.

Before the computation of credible intervals, it is crucial to check the convergence of the MCMC chains, each corresponding to one of the model parameters. This step has been performed using the R package `coda` (Plummer et al. 2006) which provides diagnostic functions to test for convergence (e.g. Geweke's test), crosscorrelations, and autocorrelations. All the results suggested a good degree of mixing and convergence of the MCMC chains.

We first report in Fig. 1a the resulting graph estimate. As described in Sect. 2.2, our method for structure learning is based on the computation of a credible interval (here at the 90% level) for each of the model parameters. Figure 1b summarizes the credible intervals and quantiles of parameters involving variables that were estimated

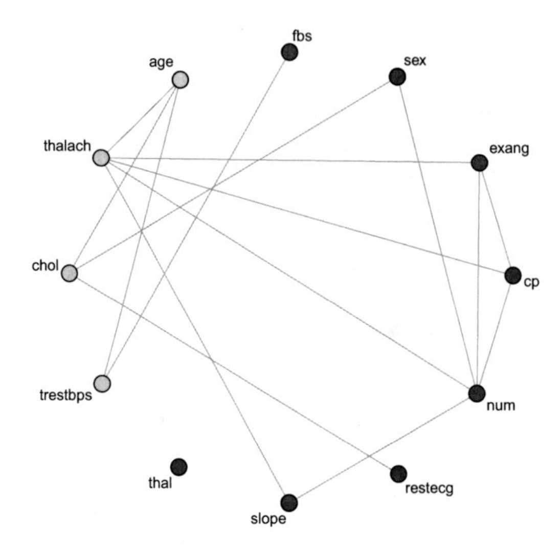

(a) Estimated graph with light (dark) grey dots representing continuous (categorical) variables

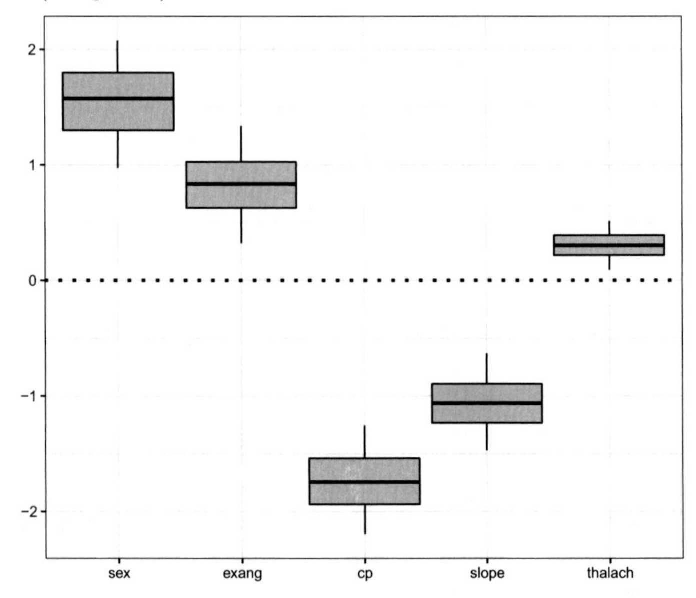

(b) Credible intervals at 90% posterior probability and quantiles of parameters corresponding to active interaction terms with the variable *num* (diagnosis of heart disease).

Fig. 1 Heart disease data analysis: Results

to be connected with "diagnosis of heart disease" (*num*). All of them do not include the zero value in their 90% credible sets, and accordingly, the corresponding edges are included in the graph estimate of Fig. 1a. Similar results, which we do not include for brevity, were obtained for the other parameters.

4 Conclusion and Next Steps

In this work, we discussed a new Bayesian methodology which allows to jointly model conditional independencies among mixed variables and perform structure learning of related undirected graphs. The novelty of our work is the adoption of the Conditional Gaussian (CG) setting of Lauritzen and Wermuth (1989) to provide an extension of existing Bayesian methodologies which deal with Gaussian and categorical data separately (Castelletti et al., 2018; Castelletti and Peluso, 2021), to parameter inference and structure learning of mixed data.

The proposed framework can suffer from scalability with respect to the computation of the normalizing constant λ_0 in the model formulation based on the canonical representation. This is due to the sum over the domain $\mathcal{Z}$ which increases exponentially in the number of categorical variables, and the computation of the inverse matrix $\mathbf{K}^{-1}$. Scalability is an issue also in the optimization procedure where the Hessian matrix must be computed at each step of the algorithm. Accordingly, we are currently investigating alternative strategies for the MCMC initialization. Furthermore, our approach for graph estimation requires the choice of the α value to build credible intervals; different choices of α might influence edge inclusion, thus leading to graph estimates with different degrees of sparsity. Sensitivity analyses can be performed when the method is applied to real data, for instance by varying the α level in a grid of distinct values. Alternative frequentist methods, such as Cheng et al. (2017), are also affected by the choice of tuning parameters that are related to confidence levels for model parameters and to penalty terms for the likelihood. Simulation experiments in which we compare our method with the approach of Cheng et al. (2017) are under investigation; preliminary studies show promising results both in terms of parameter estimation and graph recovery.

Further applications of the proposed methodology can be related to the nanotechnologies industry. In particular, Cadmium Selenide is found to exhibit the one-dimensional morphologies of nanowires, nanobelts, and nanosaws Ma & Wang (2005), often with the three morphologies intimately related. Statistically, nanostructure data are represented as categorical data corresponding to the dominant type of nanostructures. The probability of obtaining each nanostructure is expressed as a function of the predictor variables that, due to inner noise, are themselves random variables. The three key process variables affecting the morphology of the nanostructures are the continuous variables temperature, pressure, and the distance from the source material of the substrate where the deposition of nanostructures is collected (Dasgupta et al. 2008). Our framework can therefore rigorously analyse the dependence among the three nanostructures, learn how process variables can affect nanostructure occurrence probabilities, and find the experimental setting leading to the maximum probability of a given morphology.

References

Bhadra, A., Rao, A., & Baladandayuthapani, V. (2018). Inferring network structure in non-normal and mixed discrete-continuous genomic data: Inferring Network Structure in Non-Normal and Mixed Discrete-Continuous Genomic Data. *Biometrika, 74*(1), 185–195.

Bawendi, M. G., Kortan, A. R., Steigerwald, M. L., & Brus, L. E. (1989). X-Ray structural characterization of larger cadmium selenide (CdSe) semiconductor clusters. *Journal of Chemical Physics, 111*, 2564–2571.

Castelletti, F., Consonni, G., Della Vedova, M., & Peluso, S. (2018). Learning Markov equivalence classes of directed acyclic graphs: An objective Bayes approach. *Bayesian Analysis, 13*(4), 1235–1260.

Castelletti, F., & Peluso, S. (2021). Equivalence class selection of categorical graphical models. *Computational Statistics and Data Analysis, 164*, 107304.

Chen, S., Witten, D., & Shojaie, A. (2015). Selection and estimation for mixed graphical models. *Biometrika, 102*(1), 47–64.

Cheng, J., Tianxi, L., Levina, E., & Zhu, J. (2017). High-Dimensional mixed graphical models. *Journal of Computational and Graphical Statistics, 26*(2), 367–78.

Dasgupta, T., Ma, C., Joseph, V. R., Wang, Z. L., & Wu, C. J. (2008). Statistical modeling and analysis for robust synthesis of nanostructures. *Journal of the American Statistical Association, 103*(482), 594–603.

Degroot, M. (2004). *Optimal statistical decisions*. Wiley.

Frydenberg, M., & Lauritzen, S. (1989). Decomposition of maximum likelihood in mixed graphical interaction models. *Biometrika, 76*(3), 539–555.

Hodes, G., Albu-Yaron, A., Decker, F., & Motisuke, P. (1987). Three-Dimensional quantum-size effect in chemically deposited cadmium selenide films. *Physics Review B, 36*, 4215–4221.

Kalisch, M., & Bühlmann, P. (2007). Estimating high-dimensional directed acyclic graphs with the PC-Algorithm. *Journal of Machine Learning Research, 8*(22), 613–636.

Lauritzen, S. (1996). *Graphical models*. Oxford Press

Lauritzen, S., & Wermuth, N. (1989). Graphical models for associations between variables, some of which are qualitative and some quantitative. *The Annals of Statistics, 17*(1), 31–57.

Lee, J., & Hastie, T. (2015). Learning the structure of mixed graphical models. *Journal of Computational and Graphical Statistics, 24*(1), 230–53.

Ma, C., Ding, Y., Moore, D. F., Wang, X., & Wang, Z. L. (2004). Single-Crystal CdSe nanosaws. *Journal of the American Chemical Society, 126*, 708–709.

Ma, C., & Wang, Z. L. (2005). Roadmap for controlled synthesis of CdSe nanowires, nanobelts and nanosaws. *Advanced Materials, 17*, 1–6.

Martin, A. D., Quinn, K. M., & Park, J. H. (2011). Markov Chain Monte Carlo in R. *Journal of Statistical Software, 42*(9), 1–21.

Meinshausen, N., & Bühlmann, P. (2006). High-dimensional graphs and variable selection with the Lasso. *The Annals of Statistics, 34*(3), 1436–1462.

Plummer, M., Best, N., Cowles, K., & Vines, K. (2006). CODA: Convergence diagnosis and output analysis for MCMC. *R News, 6*(1), 7–11.

Yang, E., Allen, G., Liu, Z., & Ravikumar, P. (2012). Graphical models via generalized linear models. *Advances in Neural Information Processing Systems, 25*.

Yang, E., Baker, Y., Ravikumar, P., Allen, G., & Liu, Z. (2014). Mixed graphical models via exponential families. *Proceedings of the Seventeenth International Conference on Artificial Intelligence and Statistics, 33*, 1042–1050.

Zareifard, H., Rezaei Tabar, V., & Plewczynski, D. (2021). A gibbs sampler for learning DAG: A unification for discrete and gaussian domains. *Journal of Statistical Computation and Simulation, 91*(14), 2833–53.

Marginals Matrix Under a Generalized Mallows Model Based on the Power Divergence

Maria Kateri and Nikolay I. Nikolov

Abstract In rank data modeling, the classical distance-based approach has been extended by many authors in order to capture certain ranking features. Here, we study the effect on the Marginals matrix by generalizing the Mallows model via the Cressie-Read power divergence. Marginals matrices, induced by different ranking models, are extensively compared and their symmetric property over the class of Hoeffding distances is rigorously proved. Furthermore, the presented results are illustrated on a real data example for several special distances and models.

Keywords Bistochastic matrices · Distances on permutations · ϕ-divergence measures · Rank data models

1 Introduction

Rank data can be found in a wide range of scientific fields that includes psychology, biology, sociology, and economics. Such data usually arise when a set of individuals or objects needs to be ranked in accordance with some criterion. In general, rankings can be classified into two types: complete and partial, depending on whether it is required to rank all items or not. Here, we restrict our attention to the analysis of complete rank data which frequently is observed in problems with a relatively small number of ranked objects, most commonly between 3 and 10.

One of the most widely used probabilistic models for rankings is the Mallows model (MM); cf. Diaconis (1988), Fligner and Verducci (1986), and Mallows (1957). It is formulated by using distances on permutations and captures the main structure of the data with only one parameter. This makes MM a convenient initial tool for analyzing rank data and often it is used as a basis for further research. There is a

M. Kateri (✉) · N. I. Nikolov
Institute of Statistics, RWTH Aachen University, 52056 Aachen, Germany
e-mail: maria.kateri@rwth-aachen.de

N. I. Nikolov
e-mail: Nikolay.Nikolov@rwth-aachen.de

L. Grilli et al. (eds.), *Statistical Models and Methods for Data Science*, Studies in Classification, Data Analysis, and Knowledge Organization,
https://doi.org/10.1007/978-3-031-30164-3_6

rich body of work on generalizations of MM; see Critchlow (1985), Diaconis (1988, 1989), and Marden (1995) for more details. Among them is the Marginals model in Verducci (1982), which is designed to explain the Marginals matrix, associated with the rank distribution over the ordered objects. The Marginals matrix finds applications in various practical problems and statistical techniques, for example, in the error probability model for testing the imperfect ranking in the ranked set sampling (RSS) scheme; see Chen et al. (2003) and Nikolov and Stoimenova (2020). Furthermore, the Marginals matrix plays a key role in comparing the ranking abilities of two judges or ranking methods.

Another generalization of MM is recently proposed by Kateri and Nikolov (2022). By making use of the optimal property of MM in terms of the Kullback-Leibler divergence, they extend MM to a ϕ-divergence-based class of models. The ϕ-divergence is an important divergence measure between two probability distributions that is applied in numerous statistical procedures, such as hypothesis testing (cf. Cressie and Read, 1984, 1988; Pardo, 2006), parameter estimation (cf. Cressie and Pardo, 2000; Forcina and Kateri, 2021), model selection (Cressie and Pardo, 2000), contingency tables analysis (Kateri and Papaioannou, 1997; Kateri and Agresti, 2007; Kateri, 2018), and logistic regression (Kateri and Agresti, 2010). In modeling, the ϕ-divergence is particularly useful for exploring the model structure in an information theoretical direction that provides a different view angle for understanding models and their role. Kateri and Nikolov (2022) considered in more detail a generalized Mallows model (GMM) induced by the Cressie-Read power divergence (Cressie and Read, 1984), a flexible parametric family in the ϕ-divergence class.

In this work, after reviewing the distance-based and Marginals models for rank data in Sect. 2, we derive in Sect. 3 certain properties of the Marginals matrix under the GMM of Kateri and Nikolov (2022). More specifically, in Sect. 3.1, its symmetrical structure over the class of Hoeffding distances is proved, while its particular form for some special distances is provided in Sect. 3.2. We illustrate the described structural features on a real data example in Sect. 4 and make some concluding comments in the final Sect. 5.

2 Modeling Rank Data

Let us consider the task of ranking the finite set of items $\mathcal{A} = \{A_1, \ldots, A_N\}$ and a rank function $\boldsymbol{\pi} : \mathcal{A} \to \{1, \ldots, N\}$, performing full ranking. Without loss of generality, we can assume that the set $\mathcal{A}$ is $\{1, \ldots, N\}$, i.e., the items are labeled with the first N positive integers. Then, the full ranking $\boldsymbol{\pi}$ can be expressed as $\boldsymbol{\pi} = \langle \pi(1), \ldots, \pi(N) \rangle$, with $\pi(i)$ being the rank of the i-th item. Moreover, $\boldsymbol{\pi}$ is an element of $\mathcal{S}_N$, the space of N-dimensional permutations. If $\pi^{-1}(j)$ denotes the item assigned with rank j, the ordering that corresponds to listing the items in their ranked order can be expressed as $\boldsymbol{\pi}^{-1} = \langle \pi^{-1}(1), \ldots, \pi^{-1}(N) \rangle$, which is an element of $\mathcal{S}_N$ as well. Additionally, the set $\mathcal{S}_N$, equipped with the composite function $\boldsymbol{\pi} \circ \boldsymbol{\sigma} = \langle \pi(\sigma(1)), \ldots \pi(\sigma(N)) \rangle$

as a binary operation for $\boldsymbol{\pi}, \boldsymbol{\sigma} \in \mathcal{S}_N$, forms a group with an identity element $\boldsymbol{e}_N = \langle 1, \ldots, N \rangle$.

2.1 Distances on Permutations

In this subsection, we examine the structural relations in $\mathcal{S}_N$ by considering distances on permutations and their properties. First, let us introduce the following definition of a distance between two rankings.

Definition 1 A function $d : \mathcal{S}_N \times \mathcal{S}_N \to \mathbb{R}$ is a distance on $\mathcal{S}_N$ if it satisfies the following:

(i) $d(\boldsymbol{\pi}, \boldsymbol{\sigma}) > 0$ for $\boldsymbol{\pi} \neq \boldsymbol{\sigma}$ and $\boldsymbol{\pi}, \boldsymbol{\sigma} \in \mathcal{S}_N$;
(ii) $d(\boldsymbol{\pi}, \boldsymbol{\pi}) = 0$ for every $\boldsymbol{\pi} \in \mathcal{S}_N$;
(iii) $d(\boldsymbol{\pi}, \boldsymbol{\sigma}) = d(\boldsymbol{\sigma}, \boldsymbol{\pi})$ for every $\boldsymbol{\pi}, \boldsymbol{\sigma} \in \mathcal{S}_N$.

The invariance properties under left or right compositions are defined below and play a fundamental role in various applications of distances on permutations.

Definition 2 A distance $d(\cdot, \cdot)$ on $\mathcal{S}_N$ is called right-invariant (*label-invariant*), if $d(\boldsymbol{\pi}, \boldsymbol{\sigma}) = d(\boldsymbol{\pi} \circ \boldsymbol{\tau}, \boldsymbol{\sigma} \circ \boldsymbol{\tau})$ for every $\boldsymbol{\pi}, \boldsymbol{\sigma}, \boldsymbol{\tau} \in \mathcal{S}_N$, and left-invariant (*rank-invariant*), if $d(\boldsymbol{\pi}, \boldsymbol{\sigma}) = d(\boldsymbol{\tau} \circ \boldsymbol{\pi}, \boldsymbol{\tau} \circ \boldsymbol{\sigma})$ for every $\boldsymbol{\pi}, \boldsymbol{\sigma}, \boldsymbol{\tau} \in \mathcal{S}_N$. If $d(\cdot, \cdot)$ is both right- and left-invariant, then $d(\cdot, \cdot)$ is a bi-invariant distance.

As pointed out by Critchlow (1985), the right-invariance is a necessary requirement in data modeling since it ensures that the distance does not depend on the objects labeling. If $d(\cdot, \cdot)$ is left-invariant, then $d(\boldsymbol{\pi}, \boldsymbol{\sigma})$ does not use the numerical values (integers from 1 to N) of $\boldsymbol{\pi}, \boldsymbol{\sigma} \in \mathcal{S}_N$, but only the way they are ordered. Examples of right-invariant distances are the commonly used Kendall's tau, Spearman's footrule, and Spearman's rho, which are also applied in the data analysis in Sect. 4. In fact, all eight distances employed there and listed in Table 1 are right-invariant. Among them, only Cayley and Hamming distances are left-invariant as well, i.e., bi-invariant. As discussed in Sect. 3.2, the bi-invariant property leads to a special design of the Marginals matrix under a distance-based model.

An important class of distances on $\mathcal{S}_N$ is defined as follows.

Definition 3 A distance $d(\cdot, \cdot)$ on $\mathcal{S}_N$ is called Hoeffding distance, if

$$d(\boldsymbol{\pi}, \boldsymbol{\sigma}) = \sum_{i=1}^{N} a\big(\pi(i), \sigma(i)\big), \quad \text{for every } \boldsymbol{\pi}, \boldsymbol{\sigma} \in \mathcal{S}_N, \tag{1}$$

where $a(\cdot, \cdot)$ is a real-valued function on $\{1, \ldots, N\} \times \{1, \ldots, N\}$ that satisfies $a(i, i) = 0$ and $a(i, j) = a(j, i)$.

Hoeffding distances form a rich family of distances on permutations that include Spearman's footrule, Spearman's rho, Hamming, and Lee distances. As it is given in Definition 3, they can be linearly decomposed over the rankings components. This feature is crucial in obtaining some asymptotic properties for large values of N, like the combinatorial central limit theorem (CCLT), formulated and proved by Hoeffding (1951). Furthermore, as we will show in Sect. 3, the linearity in Definition 3 affects the structure of the Marginals matrix induced by the distance models presented in the next subsection.

2.2 Generalized Mallows Model Based on the Power Divergence

A common approach for analyzing rank data is to construct a probability distribution over all permutations in $\mathcal{S}_N$. In some situations, it is reasonable to assume that there is a fixed central (*modal*) ranking $\pi_0 \in \mathcal{S}_N$ and the probability of observing a ranking $\pi \in \mathcal{S}_N$ decreases exponentially in increasing the distance from π to π_0. This corresponds to the classical distance-based model, widely known as *Mallows model* (MM), defined by

$$\mathrm{P}(\pi) = \mathrm{P}(\pi|\theta, \pi_0) = \exp\left(\theta d(\pi, \pi_0) - \psi_N(\theta)\right), \tag{2}$$

where $d(\cdot, \cdot)$ is a fixed right-invariant distance on $\mathcal{S}_N$, $\theta \in \mathbb{R}$ is a parameter, and $\psi_N(\theta)$ is a normalizing constant, independent of π_0. Regarding the modal permutation π_0, it can be either fixed or unknown. The special cases of MM, with $d(\cdot, \cdot)$ being Kendall's tau and Spearman's rho, were first proposed by Mallows (1957) and later generalized by Diaconis (1988). Under certain conditions, (2) is the closest to the discrete uniform model when the models discrepancy is measured in terms of the Kullback-Leibler divergence; see Diaconis (1988, pp. 175–176). More recently, Kateri and Nikolov (2022) extended the optimal property of MM by making use of the ϕ-divergence measures and proposed the following *generalized Mallows model* (GMM), induced by the Cressie-Read power divergence (Cressie and Read, 1984),

$$\mathrm{P}(\pi) = \mathrm{P}(\pi|\beta, \lambda, \pi_0) = (1 + \beta d(\pi, \pi_0))^{1/\lambda} \frac{1}{c(\beta, \lambda)}, \tag{3}$$

where $\beta, \lambda \in \mathbb{R} \setminus \{0\}$, $d(\cdot, \cdot)$ is a right-invariant distance on $\mathcal{S}_N$, $\pi_0 \in \mathcal{S}_N$ is a central permutation, and $c(\beta, \lambda) = \sum_{\sigma \in \mathcal{S}_N} (1 + \beta d(\sigma, \pi_0))^{1/\lambda}$ is a normalizing constant. The parameter λ is associated with the Cressie-Read divergence and determines the shape properties of (3), whereas β can be viewed as a scale parameter. When $\beta = 0$, $\lambda \to \infty$ or $\lambda \to -\infty$, GMM in (3) coincides with the uniform model $\mathrm{P}_0(\pi) = 1/N!$, while for $\lambda \to 0$ we obtain the exponential MM in (2), i.e., MM$\equiv$GMM for $\lambda = 0$. The special values $\lambda = -2$, $\lambda = -1$, and $\lambda = 1$ have good interpretation as well, since in these cases (3) corresponds to the optimal model under Neyman-modified X^2, modified Kullback-Leibler, and Pearsonian X^2 divergences, respectively; see Cressie and Read (1988) and Kateri and Nikolov (2022) for more details. Similar to MM, the permutation $\boldsymbol{\pi}_0$ is the location parameter for GMM and is associated with the consensus (modal) ranking in the population. More details on the estimating procedures and constraints for the unknown parameters $(\beta,\ \lambda,\ \boldsymbol{\pi}_0)$ can be found in Kateri and Nikolov (2022).

In the next subsection, we consider an alternative generalization of MM based on Hoeffding distances.

2.3 Marginals Model

The Marginals model is first proposed by Verducci (1982) under the name *quasi-independence model*. It is a member of the exponential family and is based on a more data-analytical approach. Its probability mass function is defined as

$$\mathrm{P}\left(\boldsymbol{\pi}\right) = \mathrm{P}\left(\boldsymbol{\pi} \mid \boldsymbol{\theta}\right) = \exp\left(\sum_{i=1}^{N}\sum_{j=1}^{N}\theta_i^{(j)}\mathbb{1}\left[\pi(i) = j\right] - \psi(\boldsymbol{\theta})\right), \quad \text{for } \boldsymbol{\pi} \in \mathcal{S}_N, \quad (4)$$

where $\boldsymbol{\theta} = \left\{\theta_i^{(j)}\right\}_{i,j=1}^{N}$ are N^2 real parameters, $\mathbb{1}\left[\cdot\right]$ is the indicator function and $\psi(\boldsymbol{\theta})$ is a normalizing constant. The aim of formula (4) is to explain the quantities

$$m_{ij} = \sum_{\pi(i)=j} \mathrm{P}\left(\boldsymbol{\pi} \mid \boldsymbol{\theta}\right), \ \text{for } i,\, j \in \{1, \ldots, N\},$$

where the summation is over all permutations $\boldsymbol{\pi} \in \mathcal{S}_N$, such that $\pi(i) = j$. The matrix $\mathbf{M} = \left\{m_{ij}\right\}_{i,j=1}^{N}$ is called *Marginals matrix* since its i-th row gives the theoretical marginal distribution of the ranks assigned to object i, and its j-th column gives the theoretical marginal distribution of the objects given rank j. From

$$\sum_{\boldsymbol{\pi} \in \mathcal{S}_N} \mathrm{P}\left(\boldsymbol{\pi} \mid \boldsymbol{\theta}\right) = 1,$$

it follows that $\mathbf{M}$ is a bistochastic matrix, i.e.,

$$\sum_{i=1}^{N} m_{ij} = 1 \quad \text{and} \quad \sum_{j=1}^{N} m_{ij} = 1.$$

Thus, there are only $(N-1)^2$ free parameters $\left\{\theta_i^{(j)}\right\}_{i,j=1}^{N}$ of the Marginals model. An extension of model (4) with more free parameters is proposed by Diaconis (1989) as an application of spectral analysis to permutation data.

Models of the MM class based on Hoeffding distances are specific cases of the Marginals model with additional constraints for the parameters $\boldsymbol{\theta}$. For example, model (2) induced by Spearman's footrule

$$d_F(\boldsymbol{\pi}, \boldsymbol{\sigma}) = \sum_{i=1}^{N} |\pi(i) - \sigma(i)|, \quad \text{for } \boldsymbol{\pi}, \boldsymbol{\sigma} \in \mathcal{S}_N, \tag{5}$$

coincides with model (4) when

$$\theta_i^{(j)} = \theta \,|j - \pi_0(i)|, \quad \text{for } i, j \in \{1, \ldots, N\}.$$

In order to study more precisely the fit of the proposed models, we present several goodness-of-fit measures in the next subsection.

2.4 Model Comparisons

Let us assume that the data sample consists of n full rankings denoted by $\boldsymbol{\pi}^* = (\boldsymbol{\pi}_1^*, \ldots, \boldsymbol{\pi}_n^*)$, where $\boldsymbol{\pi}_i^* \in \mathcal{S}_N$ for $i \in \{1, \ldots, n\}$. To test the fit of GMM with fixed parameter λ, we first consider the log-likelihood ratio statistic (LRS), defined as

$$LRS_\lambda = 2 \ln \left(\frac{\mathrm{L}_\lambda \left(\boldsymbol{\pi}^*, \hat{\beta}\right)}{\mathrm{L}_\lambda \left(\boldsymbol{\pi}^*, 0\right)} \right),$$

where $\hat{\beta}$ is the maximum likelihood estimator (MLE) of β, $\mathrm{L}_\lambda(\boldsymbol{\pi}^*, \beta)$ is the likelihood under (3) with fixed λ, and $\mathrm{L}_\lambda(\boldsymbol{\pi}^*, 0)$ is the likelihood under the uniform model ($\beta = 0$). Similarly, the LRS for the Marginals model is given by

$$LRS_m = 2 \ln \left(\frac{\mathrm{L}_m \left(\boldsymbol{\pi}^*, \hat{\boldsymbol{\theta}}\right)}{\mathrm{L}_m \left(\boldsymbol{\pi}^*, \mathbf{0}\right)} \right),$$

with $\mathrm{L}_m(\boldsymbol{\pi}^*, \boldsymbol{\theta})$ being the likelihood under (4) and $\mathrm{L}_m(\boldsymbol{\pi}^*, \boldsymbol{\theta})$ corresponding to the likelihood under the uniform model ($\boldsymbol{\theta} = \mathbf{0}$). The MLEs $\hat{\boldsymbol{\theta}} = \left\{\hat{\theta}_i^{(j)}\right\}_{i,j=1}^{N}$ can be obtained by using the Newton-Raphson method or by the algorithm proposed in Verducci (1986) that is based on minimum majorization decomposition.

Next, we quantify the *total nonuniformity* in the data sample π^* by introducing the TNU coefficient, considered by Marden (1995). Let $f(\pi) = \sum_{i=1}^{n} \mathbb{1}\{\pi_i = \pi\}$ be the frequency of a given ranking $\pi \in \mathcal{S}_N$, i.e., $f(\pi)$ is the number of observations in the sample that coincide with π. Then, the empirical probability for π is $f(\pi)/n$ and

$$TNU = 2 \sum_{\pi \in \mathcal{S}_N} f(\pi) \left(\ln\left(\frac{f(\pi)}{n}\right) - \ln\left(\frac{1}{N!}\right)\right),$$

as defined in Marden (1995, p. 145). Then, the *goodness-of-fit* of a model can be tested by combining the TNU and the associated LRS, i.e., the corresponding LRS_λ or LRS_m. The null hypothesis of *perfect fit* is rejected for large values of the statistic

$$GOF = TNU - LRS.$$

Marden (1995, p. 144) showed that under certain regularity conditions GOF has an asymptotic chi-square distribution (χ^2) with $N! - p$ degrees of freedom as $n \to \infty$, where p is the number of estimated unknown parameters ($p = 1$ for LRS_λ and $p = (N-1)^2$ for LRS_m). Although the asymptotic critical region for GOF is very easy to compute, GOF does not have a clear interpretation. Thus, Marden (1995, p. 144) considered the following coefficient:

$$R^2 = 1 - \frac{GOF}{TNU}, \tag{6}$$

which is similar to the multiple correlation coefficient in the linear regression and measures the percentage of nonuniformity in the data that is explained by the fitted model. Since R^2 is a simple transformation of GOF, for the example in Sect. 4 the significance of the fitted models is discussed only in terms of the R^2 values. Nevertheless, since the TNU value is also reported, the associated GOF values can easily be calculated.

Notice that for the Marginals model (4) a sufficient statistic is the *sample Marginals matrix* $\hat{\mathbf{M}} = \{\hat{m}_{ij}\}_{i,j=1}^{N}$ defined by

$$\hat{m}_{ij} = \frac{1}{n} \sum_{k=1}^{n} \mathbb{1}\left[\pi_k^*(i) = j\right], \quad \text{for } i, j \in \{1, \ldots, N\}.$$

Thus, the fitted Marginals matrix coincides with the empirical one ($\hat{\mathbf{M}}$). In contrast, the fitted Marginals matrices under MM and GMM depend on the estimated parameters and are further studied in the next section.

3 Marginals Matrix Under GMM

Let us denote by $\mathbf{M}(\beta, \lambda, N) = \{m_{ij}(\beta, \lambda, N)\}_{i,j=1}^{N}$ the Marginals matrix under GMM in (3). Without loss of generality, it can be assumed that $\boldsymbol{\pi}_0 = \boldsymbol{e}_N$, since $d(\cdot, \cdot)$ in (3) is right-invariant and varying the permutation $\boldsymbol{\pi}_0$ is equivalent to reordering the rows of the matrix. Then, the elements of $\mathbf{M}(\beta, \lambda, N)$ can be written as

$$m_{ij}(\beta, \lambda, N) = \sum_{\pi(i)=j} \mathrm{P}(\boldsymbol{\pi}|\beta, \lambda, \boldsymbol{e}_N), \quad \text{for } i, j \in \{1, \dots, N\}, \tag{7}$$

where $\mathrm{P}(\boldsymbol{\pi}|\beta, \lambda, \boldsymbol{e}_N)$ is defined in (3) for $\boldsymbol{\pi}_0 = \boldsymbol{e}_N$. Furthermore, let us consider the random variable

$$D_N = d(\boldsymbol{\pi}, \boldsymbol{e}_N), \tag{8}$$

where $\boldsymbol{\pi}$ is uniformly chosen from the set $\mathcal{S}_N$. Note that the distribution of D_N depends only on the distance $d(\cdot, \cdot)$. Hence, we can express the normalizing constant in (3) as

$$c(\beta, \lambda) = \frac{1}{N!} \mathrm{E}\left[(1 + \beta D_N)^{1/\lambda}\right],$$

with expectation $\mathrm{E}[\cdot]$ taken in respect to D_N. In a similar way, by considering the sets

$$\mathcal{S}_N^{(i,j)} = \{\boldsymbol{\pi} \in \mathcal{S}_N : \pi(i) = j\}, \quad \text{for } i, j \in \{1, \dots, N\}, \tag{9}$$

and the random variables

$$D_N^{(i,j)} = d\left(\boldsymbol{\pi}^{(i,j)}, \boldsymbol{e}_N\right), \quad \text{for } i, j \in \{1, \dots, N\}, \tag{10}$$

with $\boldsymbol{\pi}^{(i,j)}$ being uniformly chosen from the set $\mathcal{S}_N^{(i,j)}$, we obtain the following proposition.

Proposition 1 *The elements of the Marginals matrix $M(\beta, \lambda, N)$ under the generalized Mallows model* (3) *are given by*

$$m_{ij}(\beta, \lambda, N) = \frac{1}{N} \frac{\mathrm{E}\left[\left(1 + \beta D_N^{(i,j)}\right)^{1/\lambda}\right]}{\mathrm{E}\left[(1 + \beta D_N)^{1/\lambda}\right]}, \tag{11}$$

where D_N and $D_N^{(i,j)}$ are defined in (8) *and* (10), *respectively.*

It is worth to mention that the elements of the Marginals matrix under MM in (2) can be expressed in a similar form as in formula (11). However, by using the exponentiality of MM, the corresponding expectations in the numerator and the denominator of (11) are simplified to the moment generating functions of $D_N^{(i,j)}$ and D_N, respectively, at the parameter value θ in (2); see Fligner and Verducci (1986) for more details. Furthermore, for some special distances, like Hoeffding distances, the Marginals matrix under MM is symmetric and can be approximated for large values of N by applying the asymptotic normality results for the random variables D_N and $D_N^{(i,j)}$; see Marden (1995) and Nikolov and Stoimenova (2019). In the next subsection, we prove that in the class of Hoeffding distances the Marginals matrix preserves its symmetric property even under GMM.

3.1 Marginals Matrix Structure Under Hoeffding Distances

Proposition 2 *If the distance $d(\cdot,\cdot)$ used in the generalized Mallows model* (3) *is a Hoeffding distance, then the corresponding Marginals matrix is symmetric.*

Proof Let $\mathbf{M}(\beta,\lambda,N)=\left\{m_{ij}(\beta,\lambda,N)\right\}_{i,j=1}^{N}$ be the Marginals matrix under model (3) based on a Hoeffding distance $d(\cdot,\cdot)$, defined by (1). Then, it is straightforward to check that $d(\cdot,\cdot)$ is right-invariant. Indeed, for any $\boldsymbol{\pi},\boldsymbol{\sigma},\boldsymbol{\tau}\in\mathcal{S}_N$,

$$d(\boldsymbol{\pi}\circ\boldsymbol{\tau},\boldsymbol{\sigma}\circ\boldsymbol{\tau})=\sum_{i=1}^{N}a\big(\pi(\tau(i)),\sigma(\tau(i))\big)=\sum_{i=1}^{N}a\big(\pi(i),\sigma(i)\big)=d(\boldsymbol{\pi},\boldsymbol{\sigma}),$$

where the middle equality is obtained by rearranging the summation terms. Therefore, from (7) we have

$$\begin{aligned} m_{ij}(\beta,\lambda,N) &= \sum_{\boldsymbol{\pi}\in\mathcal{S}_N^{(i,j)}}(1+\beta d(\boldsymbol{\pi},\boldsymbol{e}_N))^{1/\lambda}\frac{1}{c(\beta,\lambda)} \\ &= \sum_{\boldsymbol{\pi}\in\mathcal{S}_N^{(i,j)}}\left(1+\beta d\left(\boldsymbol{\pi}\circ\boldsymbol{\pi}^{-1},\boldsymbol{e}_N\circ\boldsymbol{\pi}^{-1}\right)\right)^{1/\lambda}\frac{1}{c(\beta,\lambda)} \\ &= \sum_{\boldsymbol{\pi}\in\mathcal{S}_N^{(i,j)}}\left(1+\beta d\left(\boldsymbol{e}_N,\boldsymbol{\pi}^{-1}\right)\right)^{1/\lambda}\frac{1}{c(\beta,\lambda)} \\ &= \sum_{\boldsymbol{\pi}\in\mathcal{S}_N^{(i,j)}}\left(1+\beta d\left(\boldsymbol{\pi}^{-1},\boldsymbol{e}_N\right)\right)^{1/\lambda}\frac{1}{c(\beta,\lambda)}, \end{aligned} \tag{12}$$

where $\mathcal{S}_N^{(i,j)}$ is defined in (9). Since, for fixed $\boldsymbol{\pi}\in\mathcal{S}_N^{(i,j)}$, the inverse permutation $\boldsymbol{\pi}^{-1}$ is an element of $\mathcal{S}_N^{(j,i)}$, there is one-to-one correspondence between the $(N-1)!$ permutations in the sets $\mathcal{S}_N^{(i,j)}$ and $\mathcal{S}_N^{(j,i)}$. Thus, from (12) it follows that

$$m_{ij}(\beta, \lambda, N) = \sum_{\sigma \in \mathcal{S}_N^{(j,i)}} (1 + \beta d(\boldsymbol{\sigma}, \boldsymbol{e}_N))^{1/\lambda} \frac{1}{c(\beta, \lambda)} = m_{ji}(\beta, \lambda, N),$$

which completes the proof. □

Similar to MM, the Marginals matrix under GMM has a further structural symmetry for some specific distances on $\mathcal{S}_N$. In the next subsection, we outline these additional symmetric properties for a few particular distances.

3.2 *Special Cases*

From the definition of $m_{ij}(\beta, \lambda, N)$ in (7) and the linear decomposition of Spearman's footrule in (5), we obtain the following relation between the elements of the Marginals matrix under GMM with d_F,

$$m_{i,j}(\beta, \lambda, N) = m_{N-i+1,N-j+1}(\beta, \lambda, N). \tag{13}$$

Similarly, it is easy to check that under GMM with Spearman's rho, defined by

$$d_R(\boldsymbol{\pi}, \boldsymbol{\sigma}) = \sum_{i=1}^{N} (\pi(i) - \sigma(i))^2, \quad \text{for } \boldsymbol{\pi}, \boldsymbol{\sigma} \in \mathcal{S}_N,$$

the elements of the Marginals matrix have the same symmetric property (13). Hence, under GMM with d_F or d_R there are $\left[\frac{N+2}{2}\right]\left(N - \left[\frac{N}{2}\right]\right)$ different elements of the Marginals matrix, with $[x]$ being the integer part of x. The structural freedom is even reduced if we use the Lee distance, given by

$$d_L(\boldsymbol{\pi}, \boldsymbol{\sigma}) = \sum_{i=1}^{N} \min(|\pi(i) - \sigma(i)|, N - |\pi(i) - \sigma(i)|), \quad \text{for } \boldsymbol{\pi}, \boldsymbol{\sigma} \in \mathcal{S}_N.$$

Analogously to the MM case studied in Nikolov and Stoimenova (2019), the Marginals elements under GMM with d_L satisfy

$$m_{i,j}(\beta, \lambda, N) = m_{N,k}(\beta, \lambda, N),$$

where $k = N - \min(|i - j|, N - |i - j|)$. Thus, there are only $\left[\frac{N+2}{2}\right]$ different elements of the matrix $M(\beta, \lambda, N)$ induced by d_L. Moreover, Marden (1995) showed that under MM based on a bi-invariant distance the Marginals matrix has only two different elements: diagonal and off-diagonal. In the same fashion, under GMM with a bi-invariant distance, it can be proved that

$$m_{i,j}(\beta, \lambda, N) = \begin{cases} B, & \text{for } i = j \\ \dfrac{1-B}{N-1}, & \text{for } i \neq j, \end{cases}$$

where B is a constant depending on β, λ, and N. In the special cases of Cayley and Hamming distances (bi-invariant), we have

$$B = \frac{1}{N} \frac{\mathrm{E}\left[(1+\beta D_{N-1})^{1/\lambda}\right]}{\mathrm{E}\left[(1+\beta D_N)^{1/\lambda}\right]},$$

where D_{N-1} and D_N are as in (8), cf. Nikolov and Stoimenova (2020).

From the specific results derived in this subsection, it is clear that even for some of the most widely used distances on $\mathcal{S}_N$ the Marginals matrix under both MM and GMM has a limited design freedom. The next section illustrates how these restrictions might be observed in the sample Marginals matrix as well and points out the associated effect on the fit of the suggested models.

4 Illustrative Example

Mao et al. (2013) studied the human ability to make noisy comparisons of items in ranking tasks by performing tests based on counting pseudo-randomly distributed dots in images. In the conducted experiment, each ranking involved sorting of four pictures in an increasing number of plotted dots. Here, we are focusing on one experimental setting that corresponds to the easiest task for the participants. The collected data is available at http://preflib.org (Mattei and Walsh, 2013), an online library of data sets concerning preferences, and consists of $n = 794$ full rankings for $N = 4$ types of pictures with 200, 209, 218, and 227 dots. From the problem setup, it is natural to assume that the consensus ranking π_0 equals the identity permutation $e_N = \langle 1, 2, 3, 4 \rangle$. The results of fitting MM, the Mallows model (2), and GMM, the generalized Mallows model (3), for eight of the most commonly used distances on $\mathcal{S}_N$ are given in Table 1.

Since the nonuniformity of the dots data is $TNU = 650.68$, the critical value of the R^2 coefficient in (6) for GMM with fixed λ is $R^2_{crit.} = 0.9459$ (based on χ^2 with $23 = 4! - 1$ degrees of freedom and nominal level 0.05). Hence, for $\lambda = 0$, there is no significant MM fit for all eight distances in Table 1. However, we can improve the fit by changing the distributional shape via the parameter λ. Thus, under d_F, d_R, and d_K, the associated GMM is significant for values of λ close to -1. The best fit is obtained for GMM with d_R, $\lambda = -1.2699$, $\beta = -1.7696$, and $R^2 = 0.9708$. Nevertheless, for better parameter interpretation we suggest GMM with d_R and $\lambda = -1$, which has a

Table 1 Results of fitting model (3) to the dots data with free parameter λ (GMM) and with $\lambda = 0$ (MM in (2))

		GMM			MM	
Distance	Notation	$\hat{\lambda}$	$\hat{\beta}$	$R^2_{\hat{\lambda}}$	$\hat{\theta}$	$R^2_{\lambda=0}$
Spearman's footrule	d_F	−0.8567	−1.4336	0.9512	−0.3931	0.9290
Spearman's rho	d_R	−1.2699	−1.7696	0.9708	−0.1630	0.8642
Chebyshev metric	d_M	−0.8171	−3.0273	0.9171	−0.9012	0.9003
Kendall's tau	d_K	−0.9637	−3.0381	0.9620	−0.6245	0.9127
Cayley distance	d_C	−1.0271	−4.5367	0.6873	−0.8621	0.6412
Ulam distance	d_U	−0.6101	1.2544	0.5978	−0.9625	0.5699
Hamming distance	d_H	−0.7781	−1.8300	0.6674	−0.6032	0.6513
Lee distance	d_L	−0.8831	−1.7228	0.7545	−0.4535	0.7195

similar fit ($R^2 = 0.9597$) and corresponds to the optimal model under the modified Kullback-Leibler divergence, as pointed out in Sect. 2.2.

The Marginals model (4) has insignificant fit for the dots data with statistic $R^2_m = 0.9392$ and critical value $R^2_{crit.} = 0.9616$ (based on χ^2 with $15 = 4! - 3^2$ degrees of freedom and nominal level 0.05). Therefore, model (3) performs substantially better, although model (4) has 9 free parameters versus only 2 for GMM. The reason for this lies in the structure of the sample Marginal matrix $\hat{\mathbf{M}}$. The values of $\hat{\mathbf{M}}$, together with $\mathbf{M}_R$, $\mathbf{M}_L$, and $\mathbf{M}_C$, the Marginals matrices induced by the fitted GMM in Table 1 with d_R, d_L, and d_C, respectively, are given below in percentages:

$$\hat{\mathbf{M}} = \begin{pmatrix} 51.26 & 25.57 & 12.97 & 10.20 \\ 25.57 & 39.42 & 21.79 & 13.22 \\ 14.61 & 21.79 & 38.79 & 24.81 \\ 8.56 & 13.22 & 26.45 & 51.76 \end{pmatrix}, \quad \mathbf{M}_R = \begin{pmatrix} 51.78 & 25.05 & 13.48 & 9.69 \\ 25.05 & 39.31 & 22.16 & 13.48 \\ 13.48 & 22.16 & 39.31 & 25.05 \\ 9.69 & 13.48 & 25.05 & 51.78 \end{pmatrix},$$

$$\mathbf{M}_L = \begin{pmatrix} 45.24 & 20.82 & 13.13 & 20.82 \\ 20.82 & 45.24 & 20.82 & 13.13 \\ 13.13 & 20.82 & 45.24 & 20.82 \\ 20.82 & 13.13 & 20.82 & 45.24 \end{pmatrix}, \quad \mathbf{M}_C = \begin{pmatrix} 45.48 & 18.17 & 18.17 & 18.17 \\ 18.17 & 45.48 & 18.17 & 18.17 \\ 18.17 & 18.17 & 45.48 & 18.17 \\ 18.17 & 18.17 & 18.17 & 45.48 \end{pmatrix}.$$

Clearly, the elements of $\hat{\mathbf{M}}$ follow a similar pattern to the matrix structure under Spearman's footrule and rho, described in Sect. 3.2, so it is natural that models (2) and (3) with d_F and d_R fit relatively well. In contrast, the restricted freedom of the Marginals matrices under d_L and d_C leads to poor approximation of $\hat{\mathbf{M}}$ by $\mathbf{M}_L$ and $\mathbf{M}_C$, which implies the lack of explanatory power of MM and GMM for these two distances. Since the evaluation of $\hat{\mathbf{M}}$ itself does not require much computational effort than estimating the parameters in MM, GMM, and particularly in model (4), the form of $\hat{\mathbf{M}}$ can be used as a guidance for appropriate distance and model choices. Furthermore, from the results in Table 1, we can conclude that model (2) is improved remarkably better by model (3) with only one extra parameter compared to model (4) with 8 additional parameters.

5 Concluding Remarks

As a future work, it would be interesting to consider a statistic that measures the deviance of the sample Marginals matrix from the structural design under given probability model. Moreover, a simple statistical procedure could be developed to test the significance of this measurement that would reduce the computational time and resources for estimating the unknown model parameters when N, the number of ranked items, is large. Another possible direction for the continuation of the current research is to apply the presented features of the Marginals matrix under GMM in the framework of RSS or other scheme that involves comparing the ranking abilities of two judges or ranking methods.

References

Chen, Z., Bai, Z., & Sinha, B. (2003). Ranked set sampling: Theory and applications. In *Lecture Notes in Statistics* (vol. 176). New York: Springer.

Cressie, N., & Pardo, L. (2000). Minimum ϕ-divergence estimator and hierarchical testing in log-linear models. *Statistica Sinica, 10*(3), 867–884.

Cressie, N., & Read, T. (1984). Multinomial goodness-of-fit tests. *The Journal of the Royal Statistical Society, Series B Statistical Methodology, 46*(3), 440–464.

Cressie, N., & Read, T. (1988). *Goodness of fit statistics for discrete multivariate data*. New York: Springer.

Critchlow, D. E. (1985). *Metric methods for analyzing partially ranked data*. New York: Springer.

Diaconis, P. (1988). Group representations in probability and statistics. In: *IMS Lecture Notes—Monograph Series* (vol. 11), Hayward, CA.

Diaconis, P. (1989). A generalization of spectral analysis with application to ranked data. *The Annals of Statistics, 17*(3), 949–979.

Fligner, M. A., & Verducci, J. S. (1986). Distance based ranking models. *The Journal of the Royal Statistical Society, Series B Statistical Methodology, 48*(3), 359–369.

Forcina, A., & Kateri, M. (2021). A new general class of RC association models: Estimation and main properties. *The Journal of Multivariate Analysis, 184*(3), 104741, 1–16.

Hoeffding, W. (1951). A combinatorial central limit theorem. *Annals of Mathematical Statistics, 22*(4), 558–566.

Kateri, M. (2018). ϕ-Divergence in contingency table analysis. *Entropy, 20*, 324.

Kateri, M., & Agresti, A. (2007). A class of ordinal quasi symmetry models for square contingency tables. *Statistics and Probability Letters, 77*, 598–603.

Kateri, M., & Agresti, A. (2010). A generalized regression model for a binary response. *Statistics and Probability Letters, 80*, 89–95.

Kateri, M., & Nikolov, N. I. (2022). A generalized Mallows model based on ϕ-divergence measures. *The Journal of Multivariate Analysis, 190*(104958), 1–14.
Kateri, M., & Papaioannou, T. (1997). Asymmetry models for contingency tables. *Journal of the American Statistical Association, 92*, 1124–1131.
Mallows, C. (1957). Non-null ranking models. I. *Biometrika, 44*(1), 114–130.
Mao, A., Procaccia, A. D., & Chen, Y. (2013). Better human computation through principled voting. In *Twenty-Seventh AAAI Conference on Artificial Intelligence* (pp. 1142–1148).
Marden, J. I. (1995). Analyzing and modeling rank data. In *Monographs on statistics and applied probability* (Vol. 64). London: Chapman & Hall.
Mattei, N., & Walsh, T. (2013). Preflib: A library for preferences. In: *International conference on algorithmic decision theory* (pp. 259–270). Springer. http://preflib.org.
Nikolov, N. I., & Stoimenova, E. (2019). Asymptotic properties of Lee distance. *Metrika, 82*(3), 385–408.
Nikolov, N. I., & Stoimenova, E. (2020). Mallows' models for imperfect ranking in ranked set sampling AStA–Adv. *Statistical Analysis, 104*(3), 459–484.
Pardo, L. (2006). *Statistical inference based on divergence measures*. New York: Chapman & Hall.
Verducci, J. S. (1982). *Discriminating Between Two Probabilities on the Basis of Ranked Preferences*. Technical report, Stanford University Press, Redwood City.
Verducci, J. S. (1989). Minimum majorization decomposition. *Contributions to probability and statistics* (pp. 160–173). New York: Springer.

Time Series Clustering Based on Forecast Distributions: An Empirical Analysis on Production Indices for Construction

Michele La Rocca, Francesco Giordano, and Cira Perna

Abstract This paper presents and discusses a recent proposal for clustering autoregressive nonlinear time series data in which dissimilarities are computed according to their forecast distributions. The procedure uses feedforward neural networks to approximate the original nonlinear process combined with the pair bootstrap as a resampling technique. An empirical analysis of the construction sector for 21 European countries is performed. Since the COVID-19 pandemic has hit all the countries in the last part of the observational period, the analysis also evaluates the possible different group structures due to the pandemic. The results are almost identical for different forecast horizons in the pre-COVID-19 period. On the contrary, where all countries experienced severe contractions in their economic activities, the dataset shows a diverse group structure, indicating different routes and timelines for economic recovery.

Keywords Nonlinear time series · Clustering · Pair bootstrap · Forecast distribution

1 Introduction

There has been a growing interest in time series clustering in recent decades. In this context, different approaches have been proposed in the statistical literature (see Aghabozorgi et al. 2015) for review. They rely on working directly with raw data, indirectly with features extracted from the raw data, or with models built from it.

M. La Rocca (✉) · F. Giordano · C. Perna
Department of Economics and Statistics, University of Salerno, Salerno, Italy
e-mail: larocca@unisa.it

F. Giordano
e-mail: giordano@unisa.it

C. Perna
e-mail: perna@unisa.it

L. Grilli et al. (eds.), *Statistical Models and Methods for Data Science*, Studies in Classification, Data Analysis, and Knowledge Organization,
https://doi.org/10.1007/978-3-031-30164-3_7

In the first case, the so-called raw-data-based approach, the distance/similarity measure for static data is modified with an appropriate one to account for time series structure (see D'Urso et al. 2018 for examples of this approach).

Since it is not easy to work directly with raw data that are highly noisy, the feature-based approach can be implemented. It is based on a preliminary transformation of the raw time series in a feature vector of lower dimension and then on applying a clustering algorithm. In this context, attracting proposals are based on the use of autocovariances and autocorrelations (Lafuente-Rego & Vilar 2016; D'Urso & Maharaj D'Urso and Maharaj 2009), and spectral density function (D'Urso et al. 2020).

Moreover, the model-based clustering approach considers that some model or probability distribution generates each time series. Time series are considered similar when the models identified on the individual series or the residuals of the fitted model are similar. As data generating processes, ARIMA models (Piccolo 1990), TAR models (Aslan et al. 2018), and GARCH models (D'Urso et al. 2016) have been considered, *inter alia*, in the literature.

Some recent approaches rely on the use of distance criteria which compare the forecast densities estimated by using a resampling method combined with a non-parametric kernel estimator (see Alonso et al. 2006; Vilar et al. 2010).

More recently, La Rocca et al. (2021) have proposed a novel approach for clustering nonlinear autoregressive time series based on comparing their full forecast distribution at a given forecast horizon. Rather than considering patterns throughout observations, this approach introduces information on the forecast behaviour at a specific time horizon in the clustering procedure. The procedure combines a class of neural network models to approximate the original nonlinear process with the pair bootstrap as a resampling device. Under general conditions of the data generating process, the overall clustering procedure, which depends on the implemented bootstrap scheme, can deliver consistent results. Moreover, it performs well in finite small samples for a broad class of nonlinear autoregressive processes and different types of innovations.

This paper aims to present and discuss the novel clustering approach and to address its employment in a real dataset on the production index for construction, a vital business cycle indicator. In the analysis, a set of 21 European countries has been considered. The observations span the period from January 2000 to December 2020 (the base year 2015), including last year, when the COVID-19 pandemic hit Europe. In this empirical analysis, the aim is to identify the different group structures induced by the COVID-19 pandemic by using the forecast one-step ahead distribution for January 2020 (so excluding any observations from the COVID-19 pandemic), the forecast multi-step ahead distribution for January 2021, and the forecast one-step ahead distribution for January 2021 (where all models have been trained up to December 2020).

The paper is organized as follows. Section 2 introduces and reviews the time series clustering procedure based on neural network forecast distributions. Section 3 discusses the case study on clustering monthly production indices for construction. Some remarks in Sect. 4 close the paper.

2 The Clustering Procedure

Let $\{Y_t, t \in \mathbb{Z}\}$ be a real-valued stationary stochastic process modelled as a nonlinear autoregressive (NAR) model of the form

$$Y_t = g\left(Y_{t-1}, \ldots, Y_{t-p}\right) + \varepsilon_t,$$

where $g(\cdot)$ is an unknown (possibly) nonlinear function, and $\{\varepsilon_t\}$ are *iid* error terms, with $\mathbb{E}[\varepsilon_t] = 0$ and $\mathbb{E}[\varepsilon_t^2] > 0$. In the following, for the sake of simplicity, we put $\mathbf{x}'_{t-1} = (Y_{t-1}, \ldots, Y_{t-p})$.

Let $\left(\mathbf{y}^{(1)}, \ldots, \mathbf{y}^{(S)}\right)$ be S observed time series of length T generated from a DGP of the previous class, where $\mathbf{y}^{(i)} = \left(Y_1^{(i)}, \ldots, Y_T^{(i)}\right)$. The aim is to cluster time series based on their full forecast distribution at a specific future time $T + h$, with $h \geq 1$.

The proposed approach accounts for the future dynamic behaviour of the time series in the clustering procedure by using the L^r-norm distance

$$D_{r,ij} = \int \left| F^i_{T+h|T}(y) - F^j_{T+h|T}(y) \right|^r dy \quad r = 1, 2 \tag{1}$$

where $F^i_{T+h|T}(\cdot)$, $i = 1, \ldots, S$ is the forecast distribution function at a given future point $T + h$, of the series $\mathbf{y}^{(i)}$, conditioned on the information set available up to time T. Since the L^r-norm distance previously defined cannot be computed directly, La Rocca et al. (2021) have proposed a strategy in which the unknown distributions are consistently estimated by using a feedforward neural network estimator and the pair bootstrap approach. In particular, given the forecast horizon h, the unknown function $g(\cdot)$ can be approximated by using the network

$$f_{mh}\left(\mathbf{x}_{t-h}; \theta\right) = \sum_{k=1}^{m} c_k \psi\left(\mathbf{w}'_k \mathbf{x}_{t-h} + w_{k0}\right) + c_0 \tag{2}$$

with $\theta = (c_0, c_1, \ldots, c_m, \mathbf{w}_1, \ldots, \mathbf{w}_m, w_{10}, \ldots, w_{m0})$, where m is the hidden layer size; $\mathbf{w}_k$ are the vectors of weights for the connections between input layer and hidden layer, c_k, $k = 1, \ldots; m$ are the weights of the link between the hidden layer and the output; w_{k0} and c_0 are the bias terms; $\psi(\cdot)$ is a proper chosen activation function and $\mathbf{x}'_{t-h} = (Y_{t-h}, \ldots, Y_{t-h-p+1})$.

As usual in neural network applications, we assume a sigmoidal activation function such as the logistic or the hyperbolic tangent function. In this case, single hidden layer neural networks are universal approximators in that they can arbitrarily closely approximate, in an appropriate corresponding metric, to L^1-integrable functions (Hornik 1991). Moreover, they have a good performance in forecasting and are uninfluenced by the eventual increasing dimension of the time series.

The general proposed procedure is reported in the following algorithm.

Algorithm

1: Fix the forecast horizon $h \geq 1$. Let $\mathcal{X} = \{(Y_t, \mathbf{x}'_{t-h}), t = p+h, \ldots, T\}$.
2: Fix the hidden layer size m and the lag structure p and estimate the weights of the network as $\hat{\theta}_h = \arg\min_\theta \frac{1}{T-p-h+1} \sum_{t=p+h}^{T} (Y_t - f_{mh}(\mathbf{x}_{t-h}; \theta))^2$.
3: Compute the residuals from the estimated network defined as $\hat{\varepsilon}_t = Y_t - f_{mh}\left(\mathbf{x}_{t-h}; \hat{\theta}_h\right)$.
4: Compute the centred residuals: $\tilde{\varepsilon}_t = \hat{\varepsilon}_t - \frac{1}{T-p-h+1} \sum_{t=p+h}^{T} \hat{\varepsilon}_t$.
5: Resample $\{(Y_t^*, \mathbf{x}'^*_{t-h}) = (Y_t^*, Y_{t-h}^*, \ldots, Y_{t-h-p+1}^*), t = p+h, \ldots, T\}$, as an iid sample from the set of tuples $\mathcal{X}$.
6: Get the bootstrap estimate of the neural network weights:
$\hat{\theta}_h^* = \arg\min_\theta \frac{1}{T-p-h+1} \sum_{t=p+h}^{T} \left(Y_t^* - f_{mh}\left(\mathbf{x}_{t-h}^*; \theta\right)\right)^2$.
7: Compute $\hat{Y}_{T+h}^* = f_{mh}\left(Y_T, Y_{T-1}, \ldots, Y_{T-p+1}; \hat{\theta}_h^*\right) + \varepsilon_{T+h}^*$ where ε_{T+h}^* is a random sample from the centred residuals $\{\tilde{\varepsilon}_t\}$.
8: The bootstrap forecast distribution $F_{T+h|T}^*$ is given by the law of $\hat{Y}_{T+h}^*$ conditioned on $\mathcal{X}$.

As usual, the bootstrap distribution can be approximated by Monte Carlo simulations repeating B times Steps 5–7, and then computing the empirical cumulative distribution function (ECDF) of $\hat{Y}_{T+h}^b$, $b = 1, 2, \ldots, B$:

$$\hat{F}_{T+h|T}^*(y) = \frac{1}{B} \sum_{b=1}^{B} \mathbb{I}\left(\hat{Y}_{T+h}^b \leq y\right) \tag{3}$$

with $\mathbb{I}(\cdot)$ denoting the indicator function.

Some comments and remarks on some of the steps of the algorithm are in order.

In Step 1, the forecast horizon is fixed and depends on the problem at hand. It is left to the researcher.

In Step 2, selecting a suitable neural network topology is a critical issue, but it has been deeply studied from a statistical perspective. Several solutions have been proposed involving information criteria (Kuan and White 1994), pruning, stopped training and regularization (Reed 1993), and inferential techniques (Anders & Korn 1999; La Rocca & Perna 2005). Finally, it is worthwhile to stress that, although the parameters of a neural network model are unidentifiable, when focusing on prediction, the problem disappears (Hwang and Ding 1997). This latter property justifies the use of a feedforward neural network in a procedure of clustering nonlinear processes based on forecast distributions. Here, we have used an almost automatic approach based on time series cross-validation as described in Bergmeir et al. (2018).

In Step 5, the pair bootstrap has been implemented as a resampling scheme. This choice appears appropriate in neural networks since it is robust for misspecified models, as the neural networks intrinsically are.

In Step 6, the optimization problem has been treated as an NLS problem and solved with a BFGS algorithm.

In Step 7, a direct multi-step forecasting approach is considered. A separate neural network model is estimated for each forecasting horizon, and forecasts are computed only conditioning on the observed data. This choice is justified by the nonlinearity of the data generating process. Moreover, although the direct strategy may have greater variability concerning the recursive method, it is most beneficial when the model is misspecified (Chevillon 2007), a peculiar aspect of neural networks which are inherently misspecified, as we have previously pointed out.

Under very general assumptions, concerning essentially the stationarity and the α-mixing condition of the data generating process (see La Rocca et al. 2021 for details), the proposed clustering procedure can deliver consistent results in nonlinear autoregressive models. Moreover, it gives good results also in finite sample sizes for a broad class of nonlinear autoregressive processes and different types of innovations.

3 An Application to the European Construction Sector

The proposed procedure has been used to cluster the production index for construction (seasonally and calendar adjusted) for 21 European countries. The production index measures the building and construction industry activity. It is of great interest for evaluating the sector's contribution to the economy and is considered a critical business cycle indicator. The time series are observed from January 2000 to December 2020 (the base year is 2015). In the last year, the dataset includes the period in which the European countries experienced the COVID-19 pandemic, which has hit our most consolidated habits with severe challenges for social and economic systems. The dataset has been downloaded from Eurostat (https://appsso.eurostat.ec.europa.eu/nui/show.do?dataset=sts_copr_m&lang=en).

The time series plots reported in Fig. 1 show, for all the European countries, a structural break at the beginning of 2020, due to the well-known COVID-19 pandemic. This behaviour is more evident in the countries that have already experienced a nationwide lockdown during the pandemic beginning with more severe provisions and personal restrictions.

Moreover, the time plots highlight a clear non-stationarity in the mean of all the considered time series. To achieve stationarity, a necessary condition to implement the proposed clustering strategy, all the series have been pre-processed by applying the first-order differences. The plots of the differenced series for all the European countries are reported in Fig. 2. Since all the series are seasonally and calendar adjusted, the first difference is the only needed transformation to get stationarity. The difference plots highlight even better the severe effect of the pandemic on the construction sector, except for a few countries (Czechia, Finland, Romania, and Slovenia) that have experienced lighter restrictions on economic activity to limit the virus spread.

The results of the Teraesvirta test for neglected nonlinearity are reported in Table 1. For 11 out of 21 series, the test rejects the null linearity hypothesis at level 5%. However, note that, although artificial neural networks show clear advantages in the case of nonlinearity, they continue to be effective tools for linear relationships. That

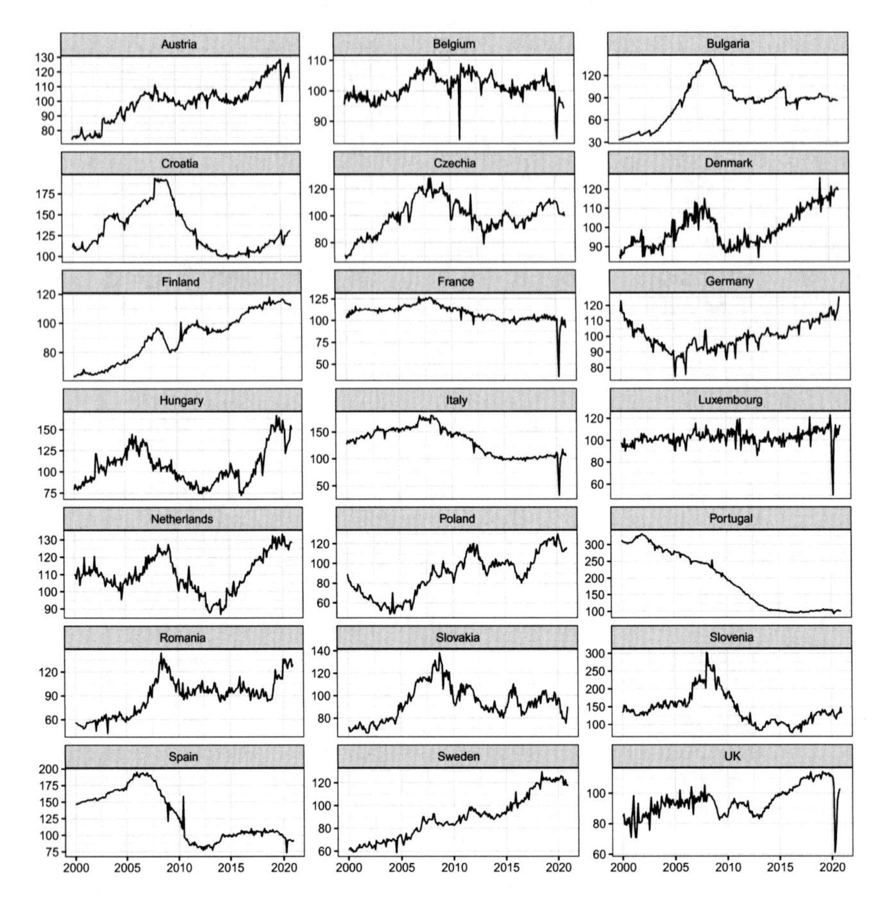

Fig. 1 Production indices for construction (base = 2015). Seasonally and calendar adjusted monthly time series from January 2000 to December 2020

allows straightforward and homogeneous modelling of both linear and nonlinear dynamics.

Neural network models have been trained by using a quadratic loss function with weight decay and a Broyden-Fletcher-Goldfarb-Shanno optimization algorithm. The authors have implemented all procedures using the R language (version 4.1.2). The bootstrap resampling scheme has been efficiently coded by exploiting the equivalence between the pair bootstrap and the multinomial weighted bootstrap and its embarrassingly parallel nature. As a result, a single bootstrap prediction distribution can be obtained with a computational burden that, on average, is in the range of tens of seconds on an Intel I7 multicore processor.

The input lag values p and the network topology's hidden layer size m to be fitted to each time series have been determined by time-series cross-validation. The results are reported in the last two columns of Table 1.

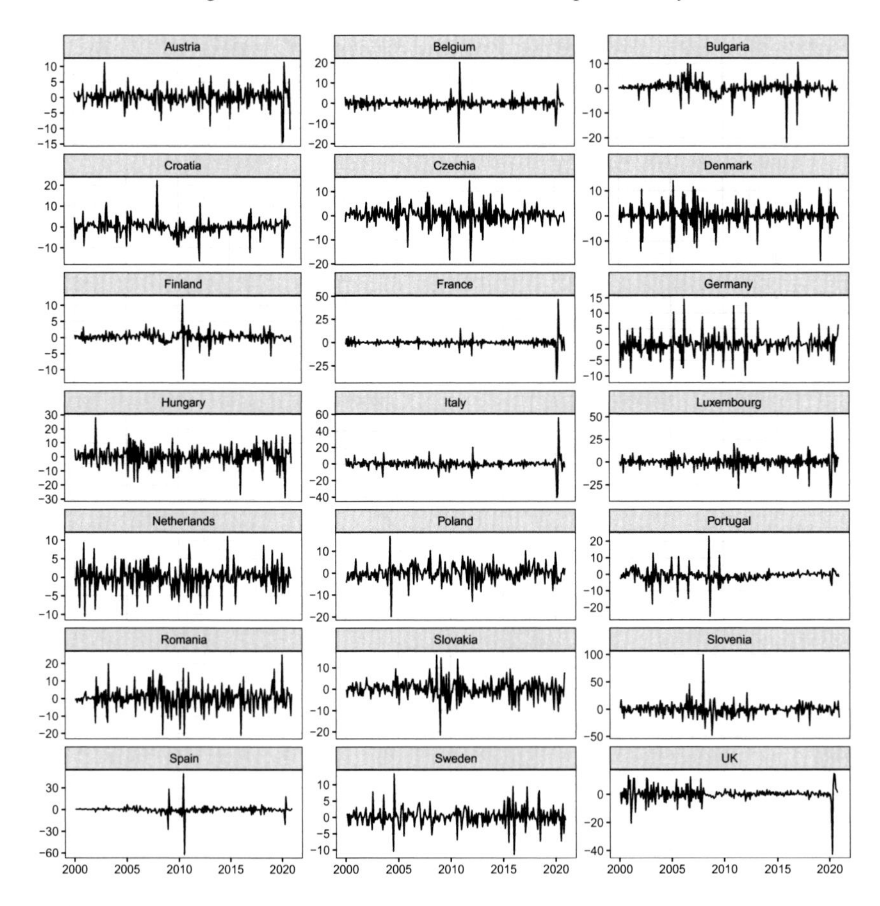

Fig. 2 First-order differences of production indices for construction (base = 2015)

To identify if the COVID-19 pandemic has induced different group structures, we have used the forecast one-step-ahead distribution for January 2020 and the forecast thirteen-step-ahead distribution for January 2021 (excluding any observations from the COVID-19 pandemic), using as training period January 2000–December 2019. Then, we added January 2019–December 2020 to the previous set (including observations from the COVID-19 pandemic) to obtain the forecast models to derive the one-step-ahead distribution for January 2021. Possible converging or diverging behaviour among the countries, as shown by the two group structures, might be due to the effect of the pandemic on the construction sector.

We have considered a hierarchical clustering technique with the average linking method. The metric is given by the bootstrap estimated counterpart of expressions (1), namely $\hat{D}_r$, with $r = 1$ and $r = 2$ estimated by the data. All bootstrap distributions have been derived using 5,000 Monte Carlo runs.

Table 1 Teraesvirta's neural network test for neglected nonlinearity (p-values, lag = 1). The values of p and m denote, respectively, the maximum lag and the hidden layer size of neural network models selected via time series cross-validation. In bold, those p-values that are lower than the threshold 0.05 (11 out of 21 time series)

	Country	p-value	p	m		Country	p-value	p	m
1	Austria	0.4865	6	4	12	Luxembourg	0.3123	5	3
2	Belgium	**0.0000**	3	3	13	Netherlands	0.3601	6	5
3	Bulgaria	0.3548	2	2	14	Poland	**0.0009**	1	8
4	Croatia	0.0560	2	6	15	Portugal	**0.0011**	6	2
5	Czechia	**0.0000**	2	5	16	Romania	0.3935	6	2
6	Denmark	0.1071	4	3	17	Slovakia	0.0644	3	2
7	Finland	**0.0000**	6	2	18	Slovenia	**0.0275**	3	3
8	France	**0.0121**	3	4	19	Spain	**0.0000**	6	10
9	Germany	**0.0019**	1	6	20	Sweden	0.8719	5	3
10	Hungary	0.3321	5	3	21	UK	**0.0005**	1	9
11	Italy	**0.0034**	2	8					

The dendrograms showing the results of the proposed clustering procedure using the L^1-norm are reported in Fig. 3a, b and c. Apparently, the group structure would have been almost identical without the impact due to the COVID-19 pandemic ($h = 1$ and $h = 13$), showing a somewhat stable economic evolution of all the countries considered in the application (see panels a and b). All groups have been identified with the average silhouette. In particular, with $h = 1$ five groups have been identified: the first with Bulgaria and Slovakia (group $\mathcal{G}_1$); the second with France, Italy, Portugal, Belgium, and Spain (group $\mathcal{G}_2$); the third with the UK, Czechia, Luxemburg, Germany, Finland, Denmark, and Poland (group $\mathcal{G}_3$); the fourth with Slovenia, Sweden, the Netherlands, Croatia, and Austria (group $\mathcal{G}_4$); Hungary constitutes a group of its own. Without the impact of the COVID-19 pandemic, the clustering would have remained the same. Of course, the countries within each group join at different distances, showing further evolution for the construction sector while preserving the group structure.

On the contrary, when the clustering is based on models that include the year 2020 in the training period, where all countries experienced severe contractions in their economic activities, the dataset shows a pretty different group structure, indicating different routes and timelines for economic recovery (see panel c). For example, Italy moved from the group $\mathcal{G}_2$ to a group that now includes Poland, Austria, Sweden, Luxembourg, and Finland. The UK moves from the group $\mathcal{G}_3$ to a group that now includes Czechia and Portugal.

In Fig. 4a, b, and c, the dendrograms with the L^2-norm are reported. Again, we have considered a hierarchical clustering technique with the average linking method and the forecast horizons $h = 1$ and $h = 13$ (with neural networks trained with observations up to December 2019) and $h = 1$ (with neural networks trained with

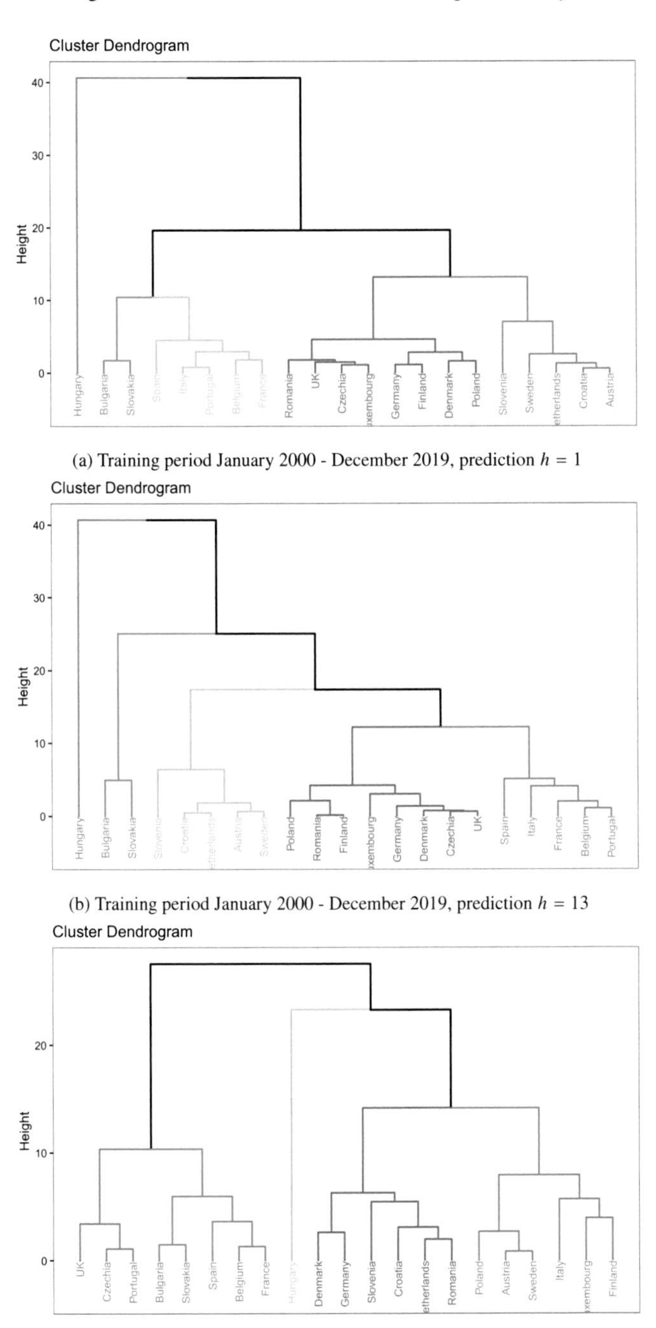

(a) Training period January 2000 - December 2019, prediction $h = 1$

(b) Training period January 2000 - December 2019, prediction $h = 13$

(c) Training period January 2000 - December 2020, prediction $h = 1$

Fig. 3 Construction index clustering based on h-step-ahead forecast distributions and L^1-norm distance

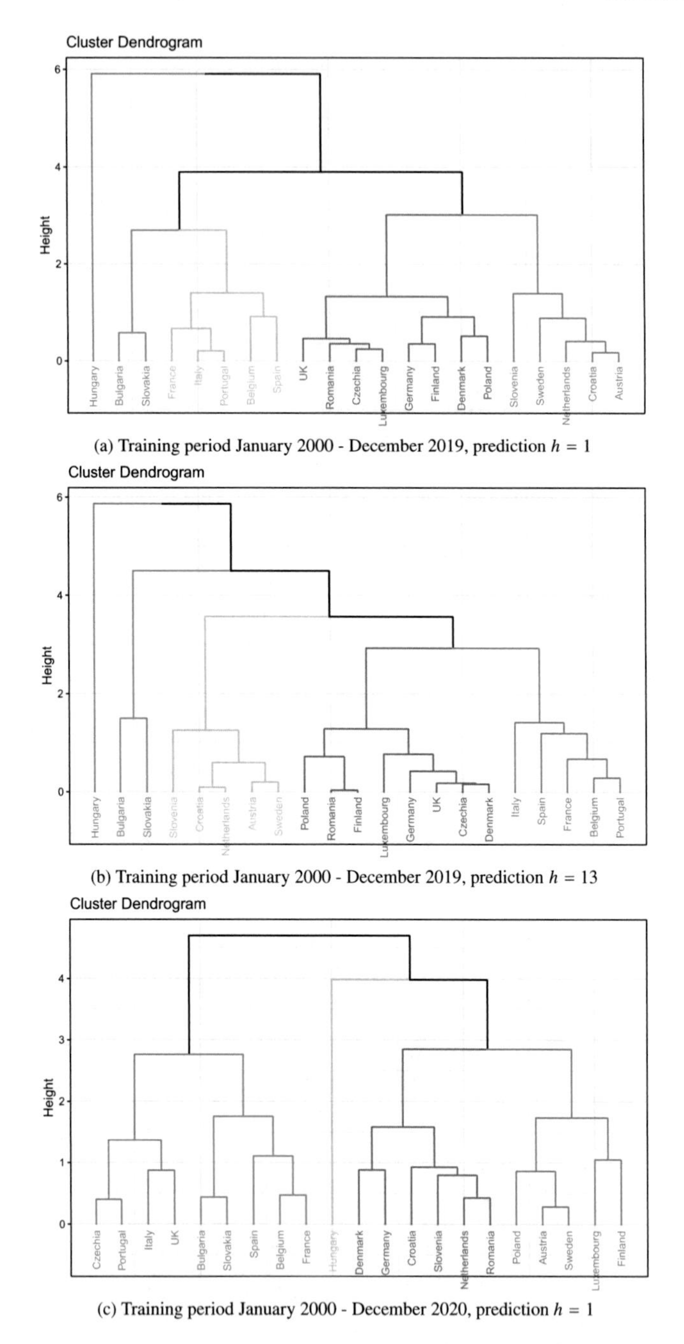

(a) Training period January 2000 - December 2019, prediction $h = 1$

(b) Training period January 2000 - December 2019, prediction $h = 13$

(c) Training period January 2000 - December 2020, prediction $h = 1$

Fig. 4 Construction index clustering based on h-step-ahead forecast distributions and L^2-norm distance

observations up to December 2020). As in the previous analysis with L^1-norm, the group structure remains almost stable with $h = 1$ and $h = 13$ (see panel Fig. 4a and b), while, once again, substantial differences are highlighted when the models include the year 2020 (see panel Fig. 4c).

Some interesting remarks are in order by comparing the results obtained with the two norms. The clustering structure seems to be not sensitive to the choice of the norm used to compute countries' distances: the groups remain almost identical in the pre-COVID-19 period. That once again shows relative stability in the dynamic evolution of the European construction sector without the COVID-19 shock. On the contrary, based on models that include the year 2020 in the training period, results obtained with the L^1-norm differ from those obtained with the L^2-norm, and different clusters of countries arise. A possible explanation might be found in the different behaviour of the two norms when structural breaks or outlying observations (both additive and innovative) are included in the training period.

4 Concluding Remarks

In this paper, we have presented and discussed an approach for clustering autoregressive nonlinear time series. It is based on dissimilarities computed according to time series forecast distributions. It uses feedforward neural networks to approximate the original process and pair bootstrap as a resampling device. An empirical analysis of the construction sector has shown the good performance of the proposed approach. Moreover, we have also empirically evaluated if the COVID-19 pandemic has affected production indices for construction by comparing the group structures pre- and post-COVID-19. In the pre-COVID period, the structure of the groups seems to be the same as the time horizon changes. In the post-COVID period, a substantial change seems to have occurred, confirming, once again, the effect that this pandemic has had on the European economy.

References

Aghabozorgi, S., Shirkhorshidi, A. S., & Wah, T. Y. (2015). Time-series clustering-a decade review. *Information Systems, 53*, 16–38.

Alonso, A. M., Berrendero, J. R., Hernández, A., & Justel, A. (2006). Time series clustering based on forecast densities. *Computational Statistics & Data Analysis, 51*(2), 762–776.

Anders, U., & Korn, O. (1999). Model selection in neural networks. *Neural Networks, 12*, 309–323.

Aslan, S., Yozgatligil, C., & Iyigun, C. (2018). Temporal clustering of time series via threshold autoregressive models: application to commodity prices. *Annals of Operations Research, 260*, 51–77.

Bergmeir, C., Hyndman, R. J., & Koo, B. (2018). A note on the validity of cross-validation for evaluating autoregressive time series prediction. *Computational Statistics & Data Analysis, 120*, 70–83.

Chevillon, G. (2007). Direct multi-step estimation and forecasting. *Journal of Economic Survey, 21*, 746–785.
D'Urso, P., & Maharaj, E. A. (2009). Autocorrelation-based fuzzy clustering of time series. *Fuzzy Sets and Systems, 160*, 3565–3589.
D'Urso, P., De Giovanni, L., & Massari, R. (2016). GARCH-based robust clustering of time series. *Fuzzy Sets and Systems, 305*, 1–28.
D'Urso, P., De Giovanni, L., & Massari, R. (2018). Robust fuzzy clustering of multivariate time trajectories. *International Journal of Approximate Reasoning, 99*, 12–38.
D'Urso, P., De Giovanni, L., Massari, R., D'Ecclesia, R. L., & Maharaj, E. A. (2020). Cepstral-based clustering of financial time series. *Expert Systems with Applications, 161*, 113705.
Hornik, K. (1991). Approximation capabilities of multilayer feedforward networks. *Neural Networks, 4*, 251–257.
Hwang, J. T. G., & Ding, A. A. (1997). Prediction intervals for artificial neural networks. *Journal of the American Statistical Association, 92*, 748–757.
Kuan, C., & White, H. (1994). Artificial neural networks: an econometric perspective. *Econometric Reviews, 13*, 1–91.
La Rocca, M., & Perna, C. (2005). Variable selection in neural network regression models with dependent data: a subsampling approach. *Computational Statistics & Data Analysis, 48*, 415–429.
La Rocca, M., Giordano, F., & Perna, C. (2021). Clustering nonlinear time series with neural network bootstrap forecast distributions. *International Journal of Approximate Reasoning, 137*, 1–15.
Lafuente-Rego, B., & Vilar, J. A. (2016). Clustering of time series using quantile autocovariances. *Advances in Data Analysis and Classification, 10*, 391–415.
Piccolo, D. (1990). A distance measure for classifying ARIMA models. *Journal of Time Series Analysis, 11*, 153–164.
Reed, R. (1993). Pruning algorithms - a survey. *IEEE Transactions on Neural Networks, 4*, 740–747.
Vilar, J. A., Alonso, A. M., & Vilar, J. M. (2010). Non-linear time series clustering based on non-parametric forecast densities. *Computational Statistics & Data Analysis, 54*(11), 2850–2865.

Partial Reconstruction of Measures from Halfspace Depth

Petra Laketa and Stanislav Nagy

Abstract The halfspace depth of a d-dimensional point x with respect to a finite (or probability) Borel measure μ in $\mathbb{R}^d$ is defined as the infimum of the μ-masses of all closed halfspaces containing x. A natural question is whether the halfspace depth, as a function of $x \in \mathbb{R}^d$, determines the measure μ completely. In general, it turns out that this is not the case, and it is possible for two different measures to have the same halfspace depth function everywhere in $\mathbb{R}^d$. In this paper, we show that despite this negative result, one can still obtain a substantial amount of information on the support and the location of the mass of μ from its halfspace depth. We illustrate our partial reconstruction procedure in an example of a non-trivial bivariate probability distribution whose atomic part is determined successfully from its halfspace depth.

Keywords Halfspace depth · Reconstruction · Characterization problem

1 The Depth Characterization/Reconstruction Problem

Let x be a point in the d-dimensional Euclidean space $\mathbb{R}^d$ and let μ be a finite Borel measure in $\mathbb{R}^d$. We write $\mathcal{H}$ for the collection of all closed halfspaces[1] in $\mathbb{R}^d$ and $\mathcal{H}(x)$ for the subset of those halfspaces from $\mathcal{H}$ that contain x in their boundary hyperplane. The *halfspace depth* (or *Tukey depth*) of the point x with respect to μ is defined as

$$D(x;\mu) = \inf_{H \in \mathcal{H}(x)} \mu(H). \tag{1}$$

[1]A halfspace is one of the two regions determined by a hyperplane in $\mathbb{R}^d$; any halfspace can be written as a set $\left\{y \in \mathbb{R}^d \colon \langle y, u\rangle \leq c\right\}$ for some $c \in \mathbb{R}$ and $u \in \mathbb{R}^d \setminus \{0\}$.

P. Laketa · S. Nagy (✉)
Faculty of Mathematics and Physics, Charles University, Prague, Czech Republic
e-mail: nagy@karlin.mff.cuni.cz

P. Laketa
e-mail: laketa@karlin.mff.cuni.cz

L. Grilli et al. (eds.), *Statistical Models and Methods for Data Science*, Studies in Classification, Data Analysis, and Knowledge Organization,
https://doi.org/10.1007/978-3-031-30164-3_8

The history of the halfspace depth in statistics goes back to the 1970s (Tukey 1975). The halfspace depth plays an important role in the theory and practice of nonparametric inference of multivariate data; for many references, see Liu et al. (1999); Nagy et al. (2019); Zuo and Serfling (2000).

The depth (1) was originally designed to serve as a multivariate generalization of the quantile function. As such, it is desirable that just as the quantile function in $\mathbb{R}$, the depth function $x \mapsto D(x;\mu)$ in $\mathbb{R}^d$ characterizes the underlying measure μ uniquely, and μ is straightforward to be retrieved from its depth. The question of whether the last two properties are valid for D is known as the *halfspace depth characterization and reconstruction problems*. They both turned out not to have an easy answer. In fact, only the recent progress in the theory of the halfspace depth gave the first definite solutions to some of these problems.

In Nagy (2021), the general characterization question for the halfspace depth was answered in the negative, by giving examples of different probability distributions in $\mathbb{R}^d$ with $d \geq 2$ with identical halfspace depth functions. On the other hand, several authors have obtained also partial positive answers to the characterization problem; for a recent overview of that work, see Nagy (2020). Only three types of distributions are known to be completely characterized by their halfspace depth functions:

(i) univariate measures, in which case the depth (1) is just a simple transform of the distribution function of μ;
(ii) atomic measures with finitely many atoms (which we subsequently call *finitely atomic measures* for brevity) in $\mathbb{R}^d$ (Struyf & Rousseeuw, 1999; Laketa & Nagy, 2021); and
(iii) measures that possess all Dupin floating bodies[2] (Nagy et al. 2019).

In this contribution, we revisit the halfspace depth reconstruction problem. We pursue a general approach and do not restrict only to atomic measures or to measures with densities. Our results are valid for any finite (or probability) Borel measure μ in $\mathbb{R}^d$. As the first step in addressing the reconstruction problem, our intention is to identify the support and the location of the atoms of μ, based on their depth. We will see at the end of this note that without additional assumptions, neither of these problems is possible to be resolved. We, however, prove several positive results.

We begin by introducing the necessary mathematical background in Sect. 2. In Sect. 3, we state our main theorem; a detailed proof of that theorem is given in the Appendix. We show that

(i) the support of the measure μ must be concentrated only in the boundaries of the level sets of its halfspace depth;
(ii) each atom of μ is an extreme point of the corresponding (closed and convex) upper level sets of the halfspace depth; and
(iii) each atom of μ induces a jump in the halfspace depth function on the line passing through that atom.

[2] A Borel measure μ on $\mathbb{R}^d$ is said to possess all Dupin floating bodies if each tangent halfspace to the halfspace depth upper level set $\left\{x \in \mathbb{R}^d : D(x;\mu) \geq \alpha\right\}$ is of μ-mass exactly α, for all $\alpha \geq 0$.

These advances enable us to identify the location of the atoms of μ, at least in simpler scenarios. We illustrate this in Sect. 4, where we give an example of a non-trivial bivariate probability measure μ whose atomic part we are able to determine from its depth. We conclude by giving an example of two measures whose depth functions are the same, yet both their supports and the location of their atoms differ.

2 Preliminaries: Flag Halfspaces and Central Regions

Notations. When writing simply a subspace of $\mathbb{R}^d$ we always mean an affine subspace, that is the set $a + L = \left\{a + x \in \mathbb{R}^d : x \in L\right\}$ for $a \in \mathbb{R}^d$ and L a linear subspace of $\mathbb{R}^d$. The intersection of all subspaces in $\mathbb{R}^d$ that contain a set $A \subseteq \mathbb{R}^d$ is called the affine hull of A, and denoted by aff (A). It is the smallest subspace that contains A. The affine hull aff $(\{x, y\})$ of two different points $x, y \in \mathbb{R}^d$ is the infinite line passing through both x and y; another example of a subspace is any hyperplane in $\mathbb{R}^d$.

For a set $A \subseteq \mathbb{R}^d$ we write $\text{int}(A)$, $\text{cl}(A)$, and $\text{bd}(A)$ to denote the interior, closure, and boundary of A, respectively. The interior, closure, and boundary of a set $B \subseteq A$ when considered only as a subset of a subspace $A \subseteq \mathbb{R}^d$ are denoted by $\text{int}_A(B)$, $\text{cl}_A(B)$, and $\text{bd}_A(B)$, respectively. For two different points $x, y \in \mathbb{R}^d$, $x \neq y$, we denote by $L(x, y)$ the interior of the line segment between x and y when considered inside the infinite line aff $(\{x, y\})$. In other words, $L(x, y)$ is the open line segment between x and y. In the special case of $A = \text{aff}\,(B)$ we write $\text{relint}(B) = \text{int}_A(B)$, $\text{relbd}(B) = \text{bd}_A(B)$, and $\text{relcl}(B) = \text{cl}_A(B)$ to denote the relative interior, relative boundary, and relative closure of B, respectively. For instance, $\text{relbd}(L(x, y)) = \{x, y\}$ and $L(x, y) = \text{relint}(L(x, y))$, but $\text{int}(L(x, y)) = \emptyset$ if $d > 1$.

We write $\mathcal{M}\left(\mathbb{R}^d\right)$ for the collection of all finite Borel measures in $\mathbb{R}^d$. For a subspace $A \subseteq \mathbb{R}^d$ and $\mu \in \mathcal{M}\left(\mathbb{R}^d\right)$ we write $\mu|_A$ to denote the measure obtained by restricting μ to the subspace A, that is the finite Borel measure given by $\mu|_A(B) = \mu(B \cap A)$ for any Borel set $B \subseteq \mathbb{R}^d$. By supp (μ) we mean the support of $\mu \in \mathcal{M}\left(\mathbb{R}^d\right)$, which is the smallest closed subset of $\mathbb{R}^d$ of full μ-mass.

2.1 *Minimizing Halfspaces and Flag Halfspaces*

For $\mu \in \mathcal{M}\left(\mathbb{R}^d\right)$ and $x \in \mathbb{R}^d$, we call $H \in \mathcal{H}(x)$ a *minimizing halfspace* of μ at x if $\mu(H) = D\,(x; \mu)$. For $d = 1$ a minimizing halfspace always trivially exists. It also exists if μ is smooth in the sense that $\mu(\text{bd}(H)) = 0$ for all $H \in \mathcal{H}(x)$, or if $\mu \in \mathcal{M}\left(\mathbb{R}^d\right)$ is finitely atomic. In general, however, the infimum in (1) does not have to be attained. We give a simple example.

Example 1 Take $\mu \in \mathcal{M}\left(\mathbb{R}^2\right)$ the sum of a uniform distribution on the disk $B = \left\{x \in \mathbb{R}^2 \colon \|x\| \leq 2\right\}$ and a Dirac measure at $a = (1, 1) \in \mathbb{R}^2$. For $x = (1, 0) \in$

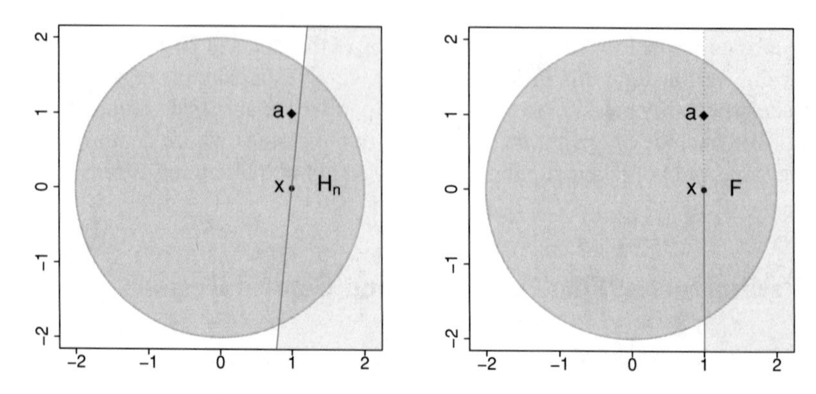

Fig. 1 The support of $\mu \in \mathcal{M}\left(\mathbb{R}^2\right)$ from Example 1 (colored disk) and its unique atom a (diamond). No minimizing halfspace of μ at $x = (1, 0) \in \mathbb{R}^2$ exists. On the left-hand panel, we see a halfspace $H_n \in \mathcal{H}(x)$ whose μ-mass is almost $D(x; \mu)$. The halfspace H_n does not contain a. On the right-hand panel, the unique minimizing flag halfspace $F \in \mathcal{F}(x)$ of μ at x is displayed

$\mathbb{R}^2$ no minimizing halfspace of μ at x exists. As can be seen in Fig. 1, the depth $D(x; \mu)$ is approached by $\mu(H_n)$ for a sequence of halfspaces $H_n \in \mathcal{H}(x)$ with inner normals $v_n = (\cos(-1/n), \sin(-1/n))$ that converge $H_v \in \mathcal{H}(x)$ with inner normal $v = (1, 0)$, yet $D(x; \mu) = \lim_{n\to\infty} \mu(H_n) < \mu(H_v)$.

For certain theoretical properties of the halfspace depth of μ to be valid, the existence of minimizing halfspaces appears to be crucial. As a way to alleviate the issue of their possible non-existence, in Pokorný et al. (2022) a novel concept of the so-called flag halfspaces was introduced. A *flag halfspace* F centered at a point $x \in \mathbb{R}^d$ is defined as any set of the form

$$F = \{x\} \cup \left(\bigcup_{i=1}^{d} \operatorname{relint}(H_i) \right). \tag{2}$$

In this formula, $H_d \in \mathcal{H}(x)$ and for each $i \in \{1, \ldots, d-1\}$, H_i stands for an i-dimensional halfspace inside the subspace $\operatorname{relbd}(H_{i+1})$ such that $x \in \operatorname{relbd}(H_i)$. The collection of all flag halfspaces in $\mathbb{R}^d$ centered at $x \in \mathbb{R}^d$ is denoted by $\mathcal{F}(x)$. An example of a flag halfspace in $\mathbb{R}^2$ is displayed in the right-hand panel of Fig. 1. That flag halfspace is a union of an open halfplane H_2 (light-colored halfplane) whose boundary passes through x, a halfline (thick halfline) in the boundary line $\operatorname{bd}(H_2)$ starting at x, and the point x itself.

The results derived from the present paper lean on the following crucial observation, whose complete proof can be found in Pokorný et al. (2022, Theorem 1).

Lemma 1 *For any $x \in \mathbb{R}^d$ and $\mu \in \mathcal{M}\left(\mathbb{R}^d\right)$ it holds true that*

$$D\left(x; \mu\right) = \min_{F \in \mathcal{F}(x)} \mu(F).$$

In particular, there always exists $F \in \mathcal{F}(x)$ such that $\mu(F) = D\left(x; \mu\right)$.

Any flag halfspace $F \in \mathcal{F}(x)$ from Lemma 1 that satisfies $\mu(F) = D\left(x; \mu\right)$ is called a *minimizing flag halfspace* of μ at x. This is because it minimizes the μ-mass among all the flag halfspaces from $\mathcal{F}(x)$. Lemma 1 tells us two important messages. First, the halfspace depth $D(x; \mu)$ can be introduced also in terms of flag halfspaces instead of the usual closed halfspaces in (1), and the two formulations are equivalent to each other. Second, in contrast to the usual minimizing halfspaces that do not exist at certain points $x \in \mathbb{R}^d$, according to Lemma 1, there always exists a minimizing flag halfspace of any μ at any x.

2.2 *Halfspace Depth Central Regions*

The upper level sets of the halfspace depth function $D(\cdot; \mu)$, given by

$$D_\alpha(\mu) = \left\{x \in \mathbb{R}^d : D\left(x; \mu\right) \geq \alpha\right\} \text{ for } \alpha \geq 0, \tag{3}$$

play the important role of multivariate quantiles in depth statistics. The set $D_\alpha(\mu)$ is called the *central region* of μ at level α. All central regions are known to be convex and closed. The sets (3) are clearly also nested, in the sense that $D_\alpha(\mu) \subseteq D_\beta(\mu)$ for $\beta \leq \alpha$. Besides (3), another collection of depth-generated sets of interest considered in Laketa and Nagy (2022); Pokorný et al. (2022) is

$$U_\alpha(\mu) = \left\{x \in \mathbb{R}^d : D\left(x; \mu\right) > \alpha\right\} \text{ for } \alpha \geq 0.$$

We conclude this collection of preliminaries with another result from Pokorný et al. (2022), which tells us that no set difference of the level sets $D_\alpha(\mu) \setminus U_\alpha(\mu)$ can contain a relatively open subset of positive μ-mass. That result lends an insight into the properties of the support of μ, based on its depth function $D\left(\cdot; \mu\right)$. It will be of great importance in the proof of our main result in Sect. 3. The complete proof of the next technical lemma can be found in Laketa et al. (2022, Lemma 3.1).

Lemma 2 *Let $\mu \in \mathcal{M}\left(\mathbb{R}^d\right)$ and let $K \subset \mathbb{R}^d$ be a relatively open set of points of equal depth of μ that contains at least two points. Then $\mu(K) = 0$.*

3 Main Result

The preliminary Lemma 2 hints that the mass of μ cannot be located in the interior of regions of constant depth. We refine and formalize that claim in the following Theorem 1, which is the main result of the present work.

In part (i) of Theorem 1 we bound the support of $\mu \in \mathcal{M}(\mathbb{R}^d)$, based on the information available in its depth function $D(\cdot; \mu)$. We do so by showing that μ may be supported only in the closure of the boundaries of the central regions $D_\alpha(\mu)$. That is a generalization of a similar result, known to be valid in the special case of finitely atomic measures $\mu \in \mathcal{M}(\mathbb{R}^d)$ (Laketa & Nagy, 2021; Liu et al. 2020; Struyf & Rousseeuw, 1999). In the latter situation, all central regions $D_\alpha(\mu)$ are convex polytopes, there is only a finite number of different polytopes in the collection $\{D_\alpha(\mu) \colon \alpha \geq 0\}$, and the atoms of μ must be located in the vertices of the polytopes from that collection. Nevertheless, not all vertices of $D_\alpha(\mu)$ are atoms of μ; an algorithmic procedure for the reconstruction of the atoms, and the determination of their μ-masses, is given in Laketa and Nagy (2021).

Extending the last observation about the possible location of atoms from finitely atomic measures to the general scenario, in part (ii) of Theorem 1 we show that all atoms of μ are contained in the extreme points[3] of the central regions $D_\alpha(\mu)$. Note that this indeed corresponds to the known theory for finitely atomic measures—the extreme points of polytopes are exactly their vertices.

Our last observation in part (iii) of Theorem 1 is that each atom $x \in \mathbb{R}^d$ of μ induces a jump discontinuity in the halfspace depth, when considered on the straight line connecting any point of higher depth with x. This will be useful in detecting possible locations of atoms for general measures.

Theorem 1 *Let* $\mu \in \mathcal{M}(\mathbb{R}^d)$.

(i) Let A be a subspace of $\mathbb{R}^d$ *that contains at least two points. Then*

$$\operatorname{supp}(\mu|_A) \subseteq \operatorname{cl}_A\left(\bigcup_{\alpha \geq 0} \operatorname{bd}_A\left(D_\alpha(\mu) \cap A\right)\right).$$

In particular, for $A = \mathbb{R}^d$ *we have*

$$\operatorname{supp}(\mu) \subseteq \operatorname{cl}\left(\bigcup_{\alpha \geq 0} \operatorname{bd}\left(D_\alpha(\mu)\right)\right).$$

(ii) Each atom x of μ *with* $D(x; \mu) = \alpha$ *is an extreme point of* $D_\beta(\mu)$ *for all* $\beta \in (\alpha - \mu(\{x\}), \alpha]$.

(iii) For any $x \in \mathbb{R}^d$ *with* $D(x; \mu) = \alpha$, *any* $z \in U_\alpha(\mu)$, *and any* $y \in \mathbb{R}^d$ *such that x belongs to the open line segment* $L(y, z)$ *between y and z, it holds true that*

[3] For a convex set $C \subset \mathbb{R}^d$, a *face* of C is a convex subset $F \subseteq C$ such that $x, y \in C$ and $(x + y)/2 \in F$ implies $x, y \in F$. If $\{z\}$ is a face of C, then z is called an *extreme point* of C.

$$D(y;\mu) \le D(x;\mu) - \mu(\{x\}).$$

The proof of Theorem 1 is given in the Appendix. Theorem 1 sheds light on the support and the location of the atoms of a measure. Its part (i) tells us that if a depth function $D(\cdot;\mu)$ attains only at most countably many different values, and each level set $D_\alpha(\mu)$ is a polytope, the mass of μ must be concentrated in the closure of the set of vertices of the level sets $D_\alpha(\mu)$. A special case is, of course, the setup of finitely atomic measures treated in Laketa and Nagy (2021); Struyf and Rousseeuw (1999).

4 Examples

We conclude this note by giving two examples. Parts (ii) and (iii) of Theorem 1 show a way, at least in special situations, to locate the atomic parts of measures. We start by reconsidering our motivating Example 1. The distribution $\mu \in \mathcal{M}(\mathbb{R}^2)$ is not purely atomic and can be shown not to possess Dupin floating bodies. Thus, it is currently unknown whether its depth function $D(\cdot;\mu)$ determines μ uniquely. In our first example of this section, we show how Theorem 1 recovers the position of the atomic part of μ. Then, in Example 3, we argue that the general problem of determining the support or the location of the atoms of $\mu \in \mathcal{M}(\mathbb{R}^d)$ from its halfspace depth is impossible to be solved without further restrictions.

Example 2 Suppose that in Example 1 we have $\mu(\{a\}) = \delta$ for $\delta \in (0, 1/2)$ small enough, and that the non-atomic part of μ is $\nu \in \mathcal{M}(\mathbb{R}^2)$ uniform on the disk B, with $\nu(B) = 1$. Hence, $\mu(\mathbb{R}^2) = \nu(B) + \mu(\{a\}) = 1 + \delta$. We first compute the halfspace depth function $D(\cdot;\mu)$ of μ and then show how to use Theorem 1 to find the atom a of μ from its depth. The computation of the depth function is done by means of determining all the central regions (3) at levels $\beta \ge 0$ of μ. We denote $\alpha = D(x;\mu)$ and split our argumentation into three situations according to the behavior of the regions $D_\beta(\mu)$:

(i) $\beta < \alpha$ where x is contained in the interior of $D_\beta(\mu)$;
(ii) $\beta \in (\alpha, \alpha + \delta]$ where x lies in the boundary of $D_\beta(\mu)$; and
(iii) $\beta > \alpha + \delta$ where $D_\beta(\mu)$ does not contain x.

First note that because ν is uniform on a unit disk, all non-empty depth regions $D_\beta(\nu)$ of ν are circular disks centered at the origin, and all the touching halfspaces[4] of $D_\beta(\nu)$ carry ν-mass exactly β.

Case I: $\beta \le \alpha$. For $\alpha = D(a;\mu) = D(a;\nu)$ we have that $D_\alpha(\nu)$ is a disk centered at the origin containing a on its boundary. Note that the halfspace depths of μ and ν remain the same outside $D_\alpha(\nu)$, since the added atom a does not lie in any minimizing halfspace of $x \notin D_\alpha(\nu)$, so we have $D_\beta(\mu) = D_\beta(\nu)$ for all $\beta \le \alpha$.

[4] We say that $H \in \mathcal{H}$ is *touching* $A \subset \mathbb{R}^d$ if $H \cap A \neq \emptyset$ and $\mathrm{int}(H) \cap A = \emptyset$.

Case II: $\beta \in (\alpha, \alpha + \delta]$. We have $D(a; \mu) = \alpha + \delta \geq \beta$, meaning that $a \in D_\beta(\mu)$. Because μ is obtained by adding mass to ν, it must be $D_\beta(\nu) \subseteq D_\beta(\mu)$ and due to the convexity of the central regions (3), the convex hull C of $D_\beta(\nu) \cup \{a\}$ must be contained in $D_\beta(\mu)$. Denote by $H \in \mathcal{H}(a)$ a touching halfspace of $D_\beta(\nu)$ that contains a on its boundary. Then $\nu(H) = \beta$, and hence $\text{int}(H) \cap D_\beta(\mu) = \emptyset$. We obtain that $D_\beta(\mu)$ is equal to the convex hull of $D_\beta(\nu) \cup \{a\}$.

Case III: $\beta > \alpha + \delta$. In a manner similar to Case II, one concludes that $D_\beta(\mu)$ is the convex hull of a circular disk $D_\beta(\nu)$ and a, intersected with the disk $D_{\beta-\delta}(\nu)$.

In order to complete the reconstruction of the atomic part of measure μ from Example 1 based on its depth function, we present Lemma 3, which is a special case of a more general result (called the *generalized inverse ray basis theorem*) whose complete proof can be found in Laketa and Nagy (2022, Lemma 4).

Lemma 3 *Suppose that $\mu \in \mathcal{M}\left(\mathbb{R}^d\right)$, $\alpha > 0$, a point $x \notin D_\alpha(\mu)$, and a face F of $D_\alpha(\mu)$ are given so that the relatively open line segment $L(x, y)$ does not intersect $D_\alpha(\mu)$ for any $y \in F$. Then there exists a touching halfspace $H \in \mathcal{H}$ of $D_\alpha(\mu)$ such that $\mu(\text{int}(H)) \leq \alpha$, $x \in H$, and $F \subset \text{bd}(H)$.*

Reconstruction. We now know the complete depth function $D(\cdot; \mu)$ of μ; see also Fig. 2. From this depth only, we will locate the atoms of μ and their mass. The only point in $\mathbb{R}^2$ that is an extreme point of more than one depth region is certainly a, so that a is the only possible candidate for an atom of μ by part (ii) of Theorem 1. Take any $\beta \in (\alpha, \alpha + \delta)$. Then $D_\beta(\mu)$ is the convex hull of a circular disk and the point a outside that disk, so its boundary contains a line segment $L(a, y_\beta)$ for $y_\beta \in \text{bd}(D_\beta(\nu))$. Due to Lemma 3, there is a halfspace $H_\beta \in \mathcal{H}$

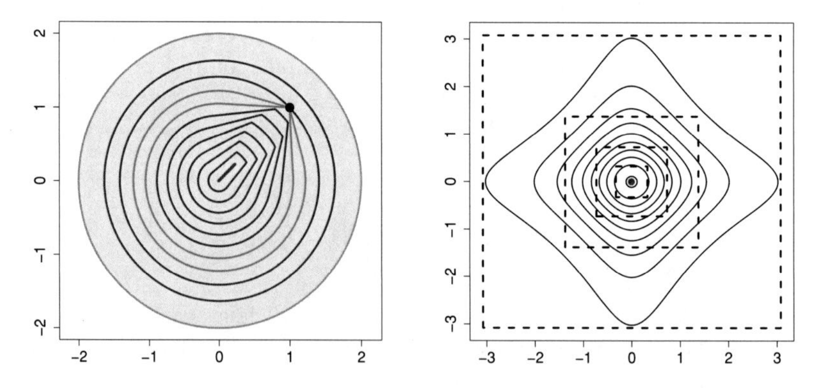

Fig. 2 Left panel: The measure μ from Example 1 being the sum of a uniform distribution on a disk and a single atom at $a \in \mathbb{R}^2$ (black point) with $\delta = 1/10$, with several contours of its depth $D(\cdot; \mu)$ (thick colored lines). The halfspace median is the red line segment in the middle of the plot. From the depth only, Theorem 1 allows us to determine the mass and the location of the atom. Two depth contours that share $a \in \mathbb{R}^2$ as an extreme point are plotted with boundaries in green. Right panel: Example 3 with $d = 2$. Several density contours of the measure $\mu \in \mathcal{M}\left(\mathbb{R}^2\right)$ (solid lines) and its atom (point at the origin), together with multiple contours of the corresponding depth $D(\cdot; \mu) \equiv D(\cdot; \nu)$ (dashed lines)

such that $L(a, y_\beta) \subset \mathrm{bd}(H_\beta)$ and $\mu(\mathrm{int}(H_\beta)) \leq \beta < \alpha + \delta = D(a; \mu) \leq \mu(H_\beta)$, the last inequality because $a \in H_\beta$. We obtain $\mu(\mathrm{bd}(H_\beta)) \geq \alpha + \delta - \beta$. This is true for any $\beta \in (\alpha, \alpha + \delta)$, and for different $\beta_1, \beta_2 \in (\alpha, \alpha + \delta)$ we have $H_{\beta_1} \neq H_{\beta_2}$ with $x \in \mathrm{bd}(H_{\beta_i})$ and $\mu\left(\mathrm{bd}(H_{\beta_i})\right) \geq \alpha + \delta - \beta_i$, $i = 1, 2$. In conclusion, we obtain uncountably many different lines $\mathrm{bd}(H_\beta)$ of positive μ-mass, all passing through a. That is possible only if a is an atom of μ, and $\mu(\{a\}) \geq \delta$. Theorem 1 again guarantees that $\mu(\{a\}) = \delta$ and that there is no other atom of μ.

The complete Example 1 gives a partial positive result toward the halfspace depth characterization problem and promises methods allowing one to determine features of μ from its depth $D(\cdot; \mu)$, at least for special sets of measures. The complete determination of the support or the atoms of μ from its depth is, however, a problem considerably more difficult, impossible to be solved in full generality. Follows an example of mutually singular measures[5] $\mu, \nu \in \mathcal{M}\left(\mathbb{R}^d\right)$ sharing the same depth function from Nagy (2020, Section 2.2).

Example 3 For $\mu_1 \in \mathcal{M}\left(\mathbb{R}^d\right)$ with independent Cauchy marginals and $\mu_2 \in \mathcal{M}\left(\mathbb{R}^d\right)$ the Dirac measure at the origin, define $\mu \in \mathcal{M}\left(\mathbb{R}^d\right)$ by the sum of μ_1 and μ_2 with weights $1/d$ and $1/2 - 1/(2d)$, respectively. The total mass of μ is hence $\mu\left(\mathbb{R}^d\right) = 1/2 + 1/(2d)$, and its support is $\mathbb{R}^d$. For the other distribution take $\nu \in \mathcal{M}\left(\mathbb{R}^d\right)$ the probability measure supported in the coordinate axes $A_i = \left\{x = (x_1, \dots, x_d) \in \mathbb{R}^d : x_j = 0 \text{ for all } j \neq i\right\}$, $i = 1, \dots, d$. The density g of ν with respect to the one-dimensional Hausdorff measure on its support $\mathrm{supp}(\nu) = \bigcup_{i=1}^d A_i$ is given as a weighted sum of densities of univariate Cauchy distributions in A_i

$$g(x) = \frac{1}{d} \sum_{i=1}^{d} \frac{\mathbb{I}[x \in A_i]}{\pi(1 + x_i^2)} \quad \text{for } x = (x_1, \dots, x_d) \in \mathbb{R}^d.$$

It can be shown (Nagy, 2020, Section 2.2) that the depths of μ and ν coincide at all points $x = (x_1, \dots, x_d)$ in $\mathbb{R}^d$

$$D(x; \mu) = D(x; \nu) = \begin{cases} \frac{1}{d}\left(\frac{1}{2} - \frac{\arctan(\max_{i=1,\dots,d}|x_i|)}{\pi}\right) & \text{if } x \in \mathbb{R}^d \setminus \{0\}, \\ 1/2 & \text{for } x = 0 \in \mathbb{R}^d. \end{cases}$$

The two measures μ and ν are, however, singular as for $A = \mathrm{supp}(\nu) \setminus \{0\}$ we have $\mu(A) = \nu(\mathbb{R}^d \setminus A) = 0$. For an arbitrary finite Borel measure, it is therefore impossible to retrieve the full information about its support only from its depth function. For a visualization of measure μ and its halfspace depth, see Fig. 2.

The same example demonstrates that in general also the location of the atoms of $\mu \in \mathcal{M}\left(\mathbb{R}^d\right)$ or even the number of them cannot be recovered from the depth function $D(\cdot; \mu)$ only—the measure ν in Example 3 does not contain any atoms, but

[5] Recall that $\mu, \nu \in \mathcal{M}\left(\mathbb{R}^d\right)$ are called *singular* if there is a Borel set $A \subset \mathbb{R}^d$ such that $\mu(A) = \nu(\mathbb{R}^d \setminus A) = 0$.

μ has a single atom at its unique halfspace median (the smallest non-empty central region (3)). Because of the very special position of the atom of μ, it is impossible to use our results from parts (ii) and (iii) of Theorem 1 to decide whether the origin is an atom of μ, or not.

5 Conclusion

The halfspace depth has many applications, for example, in classification or in non-parametric goodness-of-fit testing. However, in order to apply it properly, one needs to make sure that the measure μ is characterized by its halfspace depth function, so that we can use the halfspace depth to distinguish μ from other measures. For that reason, it is important to know which collections of measures satisfy this property. The partial reconstruction procedure provided in this paper may be used to narrow down the set of all possible measures that correspond to a given halfspace depth function. That can be used to guide the selection of an appropriate tool for depth-based analysis. The problem of determining those distributions that are uniquely characterized by their halfspace depth, however, remains open.

6 Proof of Theorem 1

For part (i), take $x \in \operatorname{supp}(\mu|_A)$ and denote $\alpha = D(x;\mu)$, $D_A = D_\alpha(\mu) \cap A$, and $U_A = U_\alpha(\mu) \cap A$. Because x comes from the support of $\mu|_A$, we know that $\mu|_A(B_x) > 0$ for any open ball B_x in A centered at x. Using Lemma 2 we conclude that B_x cannot be a subset of $D_\alpha(\mu) \setminus U_\alpha(\mu)$, meaning that it also cannot be a subset of $D_A \setminus U_A$. But, because $x \in D_A \setminus U_A$, necessarily $x \in \operatorname{bd}_A(D_A \setminus U_A) \subseteq \operatorname{bd}_A(D_A) \cup \operatorname{bd}_A(U_A)$. Now, suppose that $x \in \operatorname{bd}_A(U_A)$. Then there exists a sequence $\{x_n\}_{n=1}^\infty \subset U_A \subset A$ converging to x. We know that $\alpha_n = D(x_n;\mu) > \alpha$ for each n. Thus, for any $n = 1, 2, \dots$ we have that $x_n \in D_{\alpha_n}(\mu)$ and $x \notin D_{\alpha_n}(\mu)$, meaning that there is a point $y_n \in A$ from the set $\operatorname{bd}_A\left(D_{\alpha_n}(\mu) \cap A\right)$ in the line segment $L(x, x_n)$. Since $x_n \to x$, also the sequence $\{y_n\}_{n=1}^\infty \subset \bigcup_{\beta>\alpha} \operatorname{bd}_A\left(D_\beta(\mu) \cap A\right)$ converges to x, and necessarily $x \in \operatorname{cl}_A\left(\bigcup_{\beta>\alpha} \operatorname{bd}_A\left(D_\beta(\mu) \cap A\right)\right)$ as we intended to show.

To prove part (ii), consider $x \in \mathbb{R}^d$ such that $\mu(\{x\}) > 0$ and $\alpha = D(x;\mu)$. Choose any $y, z \in \mathbb{R}^d$ such that $x \in L(y, z)$. We will prove that one of the points y and z has depth at most $\alpha - \mu(\{x\})$, which means that x must be an extreme point of $D_\beta(\mu)$ for any $\beta \in (\alpha - \mu(\{x\}), \alpha]$. Let $F \in \mathcal{F}(x)$ be a minimizing flag halfspace of μ at x from Lemma 1, i.e. let $\mu(F) = \alpha$. We can write F in the form of the union $\{x\} \cup \left(\bigcup_{i=1}^d \operatorname{relint}(H_i)\right)$ as in (2). Since $H_d \in \mathcal{H}(x)$ is a halfspace that contains x on its boundary and x lies in the open line segment $L(y, z)$, one of the following must hold true with $j = d$:

(C_1) $L(y, z) \subset \text{relbd}(H_j)$, or
(C_2) exactly one of the points y and z is contained in $\text{relint}(H_j)$.

If (C_1) holds true with $j = d$, then we know that together with $x \in \text{relbd}(H_{d-1})$ and $x \in L(y, z)$ it implies again that one of (C_1) or (C_2) must be true with $j = d - 1$. We continue this procedure iteratively as j decreases, until we reach an index $J \in \{1, \ldots, d\}$ such that $\text{relint}(H_J)$ contains exactly one of the points y and z. Note that such an index must exist, since $\text{relbd}(H_2)$ is a halfline originating at x, so $L(y, z) \subset \text{relbd}(H_2)$ would imply either that $y \in \text{relint}(H_1)$ or that $z \in \text{relint}(H_1)$. We choose J to be the largest index $j \in \{1, \ldots, d\}$ satisfying (C_2) and assume, without loss of generality, that $y \in \text{relint}(H_J)$. Then $L(y, z) \subset \text{bd}(H_j)$ for each $j \in \{J + 1, \ldots, d\}$.

Recall that for a set $A \subset \mathbb{R}^d$ and $u \in \mathbb{R}^d$ we denote by $A + u = \{a + u : a \in A\}$ the shift of A by the vector u. Then for each $j \in \{1, \ldots, d\}$ the j-dimensional halfspace $H_j + (y - x)$ satisfies $y \in \text{relbd}(H_j + (y - x))$. Since $y \in \text{relint}(H_J)$, it must be $\text{relbd}\,(H_J + (y - x)) \subset \text{relint}(H_J)$ and therefore

$$\text{relint}\left(H_j + (y - x)\right) \subset \left(H_j + (y - x)\right) \subset \text{relint}(H_J) \text{ for } j \in \{1, \ldots, J\}, \tag{4}$$

because the relative boundaries of H_J and $H_J + (y - x)$ are parallel. At the same time, we have

$$\text{relint}\left(H_j + (y - x)\right) = \text{relint}(H_j) \text{ for each } j \in \{J + 1, \ldots, d\}, \tag{5}$$

since the indices $j \in \{J + 1, \ldots, d\}$ all satisfy $y \in \text{relbd}(H_j)$. Consider thus a shifted flag halfspace

$$F' = F + (y - x) = \{y\} \cup \bigcup_{j=1}^{d} \text{relint}\left(H_j + (y - x)\right). \tag{6}$$

Using (4), (5), and (6) we obtain

$$F' \subset \{y\} \cup \text{relint}\,(H_J) \cup \left(\bigcup_{j=J+1}^{d} \text{relint}\left(H_j\right) \right) \subset F \setminus \{x\}. \tag{7}$$

Therefore, we have $\mu(F') \leq \mu(F) - \mu(\{x\}) = \alpha - \mu(\{x\}) < \beta$ and necessarily also $D(y; \mu) < \beta$ by Lemma 1. Hence $y \notin D_\beta(\mu)$. We conclude that $x \in D_\alpha(\mu)$ cannot be contained in the relative interior of any line segment whose endpoints are both in $D_\beta(\mu)$ for $\beta \in (\alpha - \mu(\{x\}), \alpha]$, and x is therefore an extreme point of each such $D_\beta(\mu)$.

Consider now part (iii) and take F to be any minimizing flag halfspace of μ at x. Then $\mu\,(F) = D\,(x; \mu) = \alpha < D\,(z; \mu)$, and necessarily $z \notin F$. Since $x \in L(y, z)$, we can use the same argumentation as in part (ii) of this proof to conclude that exactly one of the points y and z is contained in the relative interior of one of the closed i-dimensional halfspaces H_i taking part in the decomposition of F in (2), meaning that $F \setminus \{x\}$ contains exactly one of these two endpoints y and z. Since we found that $z \notin F$, it must be that $y \in F \setminus \{x\}$. Then from (7) it follows that $F + (y - x) \subset F \setminus \{x\}$. Therefore,

$$\mu\left(F + (y - x)\right) \leq \mu\left(F\right) - \mu\left(\{x\}\right). \tag{8}$$

At the same time, Lemma 1 gives us $D(y; \mu) \leq \mu\left(F + (y - x)\right)$, which together with (8) finally implies

$$D\left(y; \mu\right) \leq \mu\left(F + (y - x)\right) \leq \mu\left(F\right) - \mu\left(\{x\}\right) = D\left(x; \mu\right) - \mu\left(\{x\}\right),$$

where the last equality follows from the fact that F is a minimizing flag halfspace of μ at x. We proved all three parts of our main theorem.

Acknowledgements The work of S. Nagy was supported by Czech Science Foundation (EXPRO project n. 19-28231X). P. Laketa was supported by the OP RDE project "International mobility of research, technical and administrative staff at the Charles University", grant CZ.02.2.69/0.0/0.0/18_053/0016976.

References

Laketa, P., & Nagy, S. (2021). Reconstruction of atomic measures from their halfspace depth. *Journal of Multivariate Analysis*, *183*, Paper No. 104727, 13. https://doi.org/10.1016/j.jmva.2021.104727.

Laketa, P., & Nagy, S. (2022). Halfspace depth for general measures: the ray basis theorem and its consequences. *Statistical Papers*, *63*(3), 849–883. https://doi.org/10.1007/s00362-021-01259-8.

Laketa, P., Pokorný, D., & Nagy, S. (2022). Simple halfspace depth. *Electronic Communications in Probability, 27*, 1–12. https://doi.org/10.1214/22-ECP503.

Liu, R. Y., Parelius, J. M., & Singh, K. (1999). Multivariate analysis by data depth: Descriptive statistics, graphics and inference. *The Annals of Statistics*, *27*(3), 783–858. http://dx.doi.org/10.1214/aos/1018031260.

Liu, X., Luo, S., & Zuo, Y. (2020). Some results on the computing of Tukey's halfspace median. *Statistical Papers*, *61*(1), 303–316. https://doi.org/10.1007/s00362-017-0941-5.

Nagy, S. (2020). The halfspace depth characterization problem. In *Nonparametric statistics*. Springer Proceedings in Mathematics & Statistics, (Vol. 339, pp. 379–389). Springer. https://doi.org/10.1007/978-3-030-57306-5_34.

Nagy, S. (2021). Halfspace depth does not characterize probability distributions. *Statistical Papers*, *62*(3), 1135–1139. https://doi.org/10.1007/s00362-019-01130-x.

Nagy, S., Schütt, C., & Werner, E. M. (2019). Halfspace depth and floating body. *Statistics Surveys*, *13*, 52–118. https://doi.org/10.1214/19-ss123.

Pokorný, D., Laketa, P., & Nagy, S. (2022). Another look at halfspace depth: Flag halfspaces with applications, under review.

Struyf, A., & Rousseeuw, P. J. (1999). Halfspace depth and regression depth characterize the empirical distribution. *Journal of Multivariate Analysis*, *69*(1), 135–153. https://doi.org/10.1006/jmva.1998.1804.

Tukey, J. W. (1975). Mathematics and the picturing of data. In: Proceedings of the International Congress of Mathematicians (Vancouver, B. C., 1974), (Vol. 2. pp. 523–531). Canadian Mathematical Congress, Montreal, Que.
Zuo, Y., & Serfling, R. (2000). General notions of statistical depth function. *The Annals of Statistics*, *28*(2), 461–482. http://dx.doi.org/10.1214/aos/1016218226.

Posterior Predictive Assessment of IRT Models via the Hellinger Distance: A Simulation Study

Mariagiulia Matteucci and Stefania Mignani

Abstract Fit assessment of item response theory models is a crucial issue. In recent years, posterior predictive model checking has become a popular tool for investigating overall model fit and potential misfit due to specific items. Different approaches rely on graphical analysis, posterior predictive p-values, the relative entropy, and, more recently, the Hellinger distance. In this study, we focus on the performance of the Hellinger distance in the case multidimensional data are analyzed with a unidimensional approach. In particular, we consider the case of three latent dimensions. A simulation study is conducted to show the effectiveness of the method. Finally, the results of an empirical application to potential three-dimensional data are discussed.

Keywords Posterior predictive model checking · Hellinger distance · IRT models · Goodness of fit

1 Introduction

Model checking is a crucial issue in statistical analysis. In the field of item response theory (IRT; see, e.g. van der Linden & Hambleton, 1997), posterior predictive model checking (PPMC; Ruben, 1984) was recently used to investigate different aspects of model fit. PPMC relies on Markov chain Monte Carlo(MCMC) estimation under a Bayesian approach, which was proved to be a very flexible method, especially for estimating models with increasing complexity. In particular, PPMC is used to assess the discrepancies between data and a model and also to identify possible limitations of the chosen model, with respect to the specific application (see Gelman et al., 2014). The PPMC method is based on the comparison between the observed and the

M. Matteucci (✉) · S. Mignani
Department of Statistical Sciences "Paolo Fortunati", University of Bologna, Bologna, Italy
e-mail: m.matteucci@unibo.it

S. Mignani
e-mail: stefania.mignani@unibo.it

L. Grilli et al. (eds.), *Statistical Models and Methods for Data Science*, Studies in Classification, Data Analysis, and Knowledge Organization,
https://doi.org/10.1007/978-3-031-30164-3_9

replicated data of a given discrepancy measure D. The main advantages of PPMC are the relative easiness to be implemented, given MCMC simulations to represent the entire posterior distribution of the parameters of interest, and the fact that the method does not require distributional assumptions, unlike classical frequentist methods. PPMC is usually implemented with graphical analyses, followed by the estimation of posterior predictive p-values (PPP-values). However, the PPP-value simply counts the number of times the replicated D is equal or higher than the realized D without addressing the magnitude of the difference between the two distributions.

Bayesian estimation via MCMC is well-established for IRT models. For this reason, also PPMC has been intensively used to examine different aspects of model fit, such as person fit or item fit, the local independence assumption, and test dimensionality (see, among others, Sinharay et al., 2006; Levy et al., 2009; Levy & Svetina, Levy and Svetina 2011). A special attention was given to the case of multidimensional data analyzed with a unidimensional approach. In fact, item response real data often show multidimensionality due to the existence of different latent variables (abilities). In these cases, estimating a unidimensional model yields a less accurate measure of individual traits as compared to fitting multiple unidimensional models with correlated traits.

To address these aspects, it was proposed to use the relative entropy (RE) within PPMC with global fit measures (Wu et al., 2014). The RE is able to measure the difference between the predictive and the realized distribution of a specific discrepancy measure and overcomes the limitations of the PPP-value, which is not able to quantify the amount of model misfit. The RE in turn suffers from two main drawbacks that may weaken its usefulness as a pairwise distance measure in applied settings: it is asymmetric and not upper bounded. For these reasons, an approach based on the Hellinger (H) distance was proposed (Matteucci & Mignani, 2021a, 2021b). In particular, the H distance was used to quantify the magnitude of the difference between the predictive and the realized distribution of measures for item pairs. In fact, similarly to the RE, the H distance provides a quantitative measure of the degree of misfit by overcoming the approach based on PPMC. Unlike the RE, the H distance satisfies the metric properties, including symmetry, and it is bounded between 0 and 1. These properties make the H distance a suitable measure for improving the interpretation of results in applied settings. Moreover, the H distance can be used for model comparison purposes, considering both single discrepancy measures on item pairs and overall. The approach based on the H distance was proposed and implemented in a under-fitting scenario to investigate the misfit of unidimensional models to multidimensional data (Matteucci & Mignani, 2021a). Also, an over-fitting scenario was considered in Matteucci and Mignani (2021b), where the fit of different multidimensional approaches to unidimensional data was evaluated.

In this study, we enlarge the study in Matteucci and Mignani (2021a) by considering IRT multidimensional models with three specific latent traits instead of two to investigate if the latent dimensionality may affect the effectiveness of the H distance to assess model fit.

The performance of the H distance is investigated for detecting the misfit of an IRT unidimensional model with both simulated and real multidimensional response data. This scenario is common especially in socio-behavioral research, where the complexity of the data structure cannot be handled with a unidimensional approach, which is more typical of the educational field. We compare our proposal to the classical PPMC-based PPP-values. The main strengths of the H distance, compared to traditional approaches, rely on the possibility to directly quantify the amount of misfit and to be used for model comparison purposes in situations with complex data structure.

The paper is organized in the following way. Section 2 briefly reviews the IRT models used in the paper. Section 3 discusses the PPMC-based approach and the discrepancy measures used for investigating local independence in IRT models. The proposal based on the H distance is also reviewed. The results of a simulation study are reported in Sect. 4, while a three-dimensional data empirical application is discussed in Sect. 5. Concluding remarks end the paper.

2 IRT Models

IRT models (see, e.g. van der Linden & Hambleton, van der Linden and Hambleton 1997), also called latent trait models, are usually employed to model the relation among categorical response variables and a set of latent continuous variables. Given a response variable vector $\mathbf{Y}_j$ containing the responses to item j, with $j = 1, ..., k$ items, given by n subjects, and a set of latent traits, also called abilities, $\boldsymbol{\theta}$, IRT models are based on the assumption of local independence, i.e.

$$P(\mathbf{Y} = \mathbf{y}|\boldsymbol{\theta}) = \prod_{j=1}^{k} P(\mathbf{Y}_j = \mathbf{y}_j|\boldsymbol{\theta}), \tag{1}$$

where $\mathbf{Y}$ collects the response variable vectors $\mathbf{Y}_j$ for all k items and $\mathbf{y}$ is its realization.

IRT models are often used in the field of educational measurement and to describe social phenomena, when the responses to a questionnaire are typically available. In fact, IRT models assume that the relations among the items may be explained by the underlying latent variables.

In its most simple formulation, an IRT model expresses the probability of a positive response to a binary item j by a subject i, with $i = 1, ..., n$ as a function of a single latent trait, as follows:

$$P(Y_{ij} = 1|\theta_i, \alpha_j, \delta_j) = \Phi(\alpha_j\theta_i - \delta_j). \tag{2}$$

Equation (2) describes the two-parameter normal ogive (2PNO) model, where α_j and δ_j represent the item j-th discrimination and difficulty, respectively. The normal ogive parameterization is chosen here instead of the most common logistic one, as it is more useful to work with Bayesian estimation via MCMC (see, e.g. Béguin & Glas, 2001).

In the case more than one latent trait is assumed, we resort to IRT multidimensional models (MIRT; Reckase (2009)). Under a confirmatory setting with m specific latent traits, with $v = 1, ..., m$ the 2PNO multi-unidimensional model (see Sheng and Wikle (2007)) is defined as follows:

$$P(Y_{vij} = 1|\theta_{vi}, \alpha_{vj}, \delta_j) = \Phi(\alpha_{vj}\theta_{vi} - \delta_j), \tag{3}$$

where α_{vj} is the v-specific item discrimination parameter for item j, δ_j is the difficulty parameter, and θ_{vi} is the v-specific latent trait for respondent i. Model (3) is based on a simple structure, i.e. each item may load on a single latent dimension only. Also, the different latent traits may be correlated.

Lastly, the additive model (see Sheng and Wikle (2009)) generalizes the multi-unidimensional model (3) by adding an overall latent trait θ_0, which is related to all the items through the discrimination parameter α_{0j}, for $j = 1, ..., k$, as follows:

$$P(Y_{vij} = 1|\theta_{0i}, \theta_{vi}, \alpha_{0j}, \alpha_{vj}, \delta_j) = \Phi(\alpha_{0j}\theta_{0i} + \alpha_{vj}\theta_{vi} - \delta_j). \tag{4}$$

According to Eq. (4), $m + 1$ latent abilities are specified, which may again be correlated. The details about Bayesian estimation via MCMC for these models can be found in Sheng and Wikle (2007) and Sheng and Wikle (2009).

3 PPMC and Discrepancy Measures for IRT Models

PPMC techniques are based on the comparison of observed data with replicated data generated or predicted by the model by using a number of diagnostic measures that are sensitive to model misfit (see Sinharay et al., 2006). Substantial differences between the posterior distribution based on observed data and the posterior predictive distribution indicate poor model fit.

Given the data $\mathbf{y}$, let $p(\mathbf{y}|\boldsymbol{\omega})$ be the likelihood for a model depending on the set of parameters $\boldsymbol{\omega}$ and $p(\boldsymbol{\omega})$ the prior distribution for the parameters, respectively. From a practical point of view, one should define a suitable discrepancy measure $D(\cdot)$ and compare the posterior distribution of $D(\mathbf{y}, \boldsymbol{\omega})$, based on observed data, to the posterior predictive distribution of $D(\mathbf{y}^{rep}, \boldsymbol{\omega})$, based on replicated data. Discrepancy measures should be chosen to capture relevant features of the data and differences among data and the model. Besides a graphical analysis, it is possible to resort to the PPP-value defined as the probability that the replicated data could be more extreme than the observed data, as measured by the test quantity (Gelman et al., 2014), as follows:

$$\text{PPP-value} = p(D(\mathbf{y}^{rep}, \boldsymbol{\omega}) \geq D(\mathbf{y}, \boldsymbol{\omega}) \mid \mathbf{y}) = \tag{5}$$

$$= \int_{D(\mathbf{y}^{rep}, \boldsymbol{\omega}) \geq D(\mathbf{y}, \boldsymbol{\omega})} p(\mathbf{y}^{rep} \mid \boldsymbol{\omega}) p(\boldsymbol{\omega} \mid \mathbf{y}) d\mathbf{y}^{rep} d\boldsymbol{\omega}. \tag{6}$$

The choice of a suitable discrepancy measure is crucial in posterior predictive assessment. In order to check for the local independence assumption in IRT models, effective diagnostic measures are based on the association or on covariance/correlation among item pairs. In this paper, we consider the model-based covariance (MBC) that depends on both data and model parameters as follows:

$$\text{MBC}_{jj'} = \frac{1}{n} \sum_{i=1}^{n} (Y_{ij} - E(Y_{ij})(Y_{ij'} - E(Y_{ij'}), \tag{7}$$

where $E(Y_{ij})$ is the expected value of the response variable depending on the specific IRT model. The MBC is found to be effective as it measures the covariance among item pairs by explicitly conditioning on the underlying latent variable. If the local independence assumption holds, the MBC is close to zero. If the local independence does not hold, the MBC is greater than zero for items loading on the same latent variable (PPP-values are close to zero) and smaller for items loading on different latent variables (PPP-values are close to one). Similar results may be obtained with the Mantel-Haenszel (MH) statistic which is defined as the odds ratio conditionally to the rest score, i.e. the raw test score obtained by excluding the particular item pair under consideration. Unlike the MBC, the MH statistics is based on data only. Lastly, in Levy and Svetina (2011) it was proposed an overall measure, namely the generalized dimensionality discrepancy measure (GDDM), defined as the mean of the absolute values of MBC over unique item pairs. The GDDM is a unidirectional measure of average conditional covariance. When the GDDM is equal to zero, a weak local independence for all the item pairs is assumed. If the assumption of local independence is violated, the GDDM is greater than zero and the PPP-value will be close to zero.

As the approach based on the PPP-value only counts the number of replications for which the predictive discrepancy exceeds the realized one, a relevant improvement in posterior predictive assessment would be to quantify the amount of the difference between the two distributions, in case the chosen discrepancy measure depends on both data and model parameters.

To this aim, the use of the RE was proposed in Wu et al. (2014). However, this approach has several limitations due to the lack of interpretability and the possibility to make comparisons in applied settings, as the RE is not upper bounded. For this reason, to quantify the difference between the realized and the predictive distribution within PPMC, in Matteucci and Mignani (2021a) it was proposed to use the H distance which is symmetric, it does obey the triangle inequality and its range is 0-1. The interpretation of the H distance follows the empirical rule "the smaller the better". The direct calculation is computationally demanding and, given the MCMC simulations, it is usually estimated by the normal kernel density. Given a discrepancy measure $D(\cdot)$, the H distance in the context of PPMC is defined as

$$\mathrm{H}(P, Q) = \sqrt{1 - \int \sqrt{p(D(\mathbf{y}, \boldsymbol{\omega}))p(D(\mathbf{y}^{rep}, \boldsymbol{\omega}))}d\mathbf{y}d\boldsymbol{\omega}}, \tag{8}$$

where P and Q are two probability distributions of continuous random variables.

In order to check for local independence, we used the H distance with the MBC discrepancy measure (MBC-H) to take into account a fit measure for each item pair and with the GDDM measure (GDDM-H) to evaluate the overall fit based on item pairs. In Matteucci and Mignani (2021a), it was proposed to investigate the assumption of local independence for 2PNO models by focusing on multidimensional data analyzed with the unidimensional model, for the case of $m = 2$ specific latent variables. Also, in Matteucci and Mignani (2021b), a study in an over-fitting scenario was proposed, where unidimensional data are analyzed through different multidimensional models.

In this paper, we focus on an under-fitting scenario, when multidimensional data are treated through a unidimensional approach, but for the case of $m = 3$ specific latent traits to study if the number of dimensions may affect the effectiveness of the H distance to assess model fit.

4 Simulation Study

In this study, we consider multidimensional data simulated from either a multi-unidimensional model with $m = 3$ specific latent abilities or an additive model with $m + 1 = 4$ latent abilities, i.e. three specific and one overall traits. The test length is established in $k = 15$, with 3 subtests of length $k_1 = k_2 = k_3 = 5$, and the sample size is fixed at $n = 1{,}000$. The item parameters are drawn from the following distributions: $\alpha_0 \sim U(1, 2)$, $\alpha_v \sim U(1, 2)$, $\delta \sim U(-2, 2)$. The latent scores are drawn from a multivariate normal $\boldsymbol{\theta} \sim MN(\mathbf{0}, \boldsymbol{\Sigma})$, where $\boldsymbol{\Sigma}$ is the correlation matrix with off-diagonal elements set equal to 0, 0.3, 0.6, and 0.9. to define different simulation conditions.

The parameters of the data-analysis model are estimated via the Gibbs sampler. A number of 5,000 MCMC iterations are conducted, where 1,000 are the burn-in iterations. The effective samples are thinned by 4 so that 1000 samples are used in PPMC. For the unidimensional model, a standard normal prior is used for the latent trait and the difficulty parameters. A standard normal distribution truncated at zero to the left is used as prior distribution for the discrimination parameters to ensure positivity. For the multi-unidimensional and the additive models, conjugate standard normal priors are specified for the item parameters, and a conjugate multivariate normal prior is used for the covariance matrix of the latent traits (see Sheng 2008b, 2010). Finally, 100 replications are done for each simulation condition. The software

Table 1 Proportion of extreme PPP-values for the MH statistic and the MBC and PPP-value for the GDDM when the same model is used to simulate and analyze data

Model	ρ	Proportion of extreme PPP-values		GDDM
		MH	MBC	PPP-value
Multi-uni	0.0	0.000	0.000	0.546
	0.3	0.000	0.000	0.529
	0.6	0.000	0.000	0.546
	0.9	0.000	0.000	0.181
Additive	0.0	0.000	0.000	0.376
	0.3	0.000	0.000	0.531
	0.6	0.000	0.000	0.491
	0.9	0.000	0.000	0.376

MATLAB was used both to estimate the model parameters and to implement PPMC and the proposed H distance.[1]

The performance of the PPP-values and the proposed MBC-H and GDDM-H at detecting misfit is evaluated. We use the proportion of extreme PPP-values (below 0.05 or above 0.95) among the item pairs to summarize the results for the MH statistic and the MBC. We report some descriptive statistics for the MBC-H and the GDDM-H.

The case of the same multidimensional model to both generate and fit the data is also considered to investigate if the chosen measures are able to report good fit in these cases. In this specific case, for all conditions, the proportion of extreme PPP-values for the MBC computed on the 105 different item pairs indicates good fit (see Table 1). Also, the PPP-values for the GDDM are all around 0.5, meaning good fit, with the exception of the case of strong correlations ($\rho = 0.9$) for the multi-unidimensional model (0.181). Even if this last PPP-values is not extreme, the deviation from 0.5 may be due to the strong correlation between the traits which make reasonable the data unidimensionality.

The results on the MBC-H and the GDDM-H are reported in Table 2. In Matteucci and Mignani (2021a), it was established 0.5 as a possible threshold for model fit. Here, the average values for MBC-H are below 0.5, indicating good fit, for all the conditions except again the case of $\rho = 0.9$ in the multi-unidimensional model. Analogously, it can be observed that the average value of the MBC-H is 0.454, close to 0.5, for the additive model with $\rho = 0.9$. Overall, the values of the MBC-H tend to increase as the correlation among the traits increases, as a multidimensional approach is progressively not needed when data become nearly unidimensional. Also, the proportion of MBC-H values greater than 0.5 is equal to zero for all the cases,

[1] MATLAB packages provided in Sheng (2008a, b, 2010) are used for data generation and for item parameter estimation. The Authors wrote MATLAB-specific programs for performing PPMC with the H distance. The documentation is still not ready but researchers interested in the scripts may ask the Authors for free.

Table 2 MBC-H and GDDM-H when the same model is used to simulate and analyze data

Model	ρ	MBC-H						GDDM-H
		Mean	Median	SD	Min	Max	Prop > 0.5	
Multi-uni	0.0	0.347	0.343	0.070	0.183	0.497	0.000	0.264
	0.3	0.349	0.348	0.052	0.180	0.456	0.000	0.282
	0.6	0.364	0.366	0.053	0.222	0.496	0.000	0.281
	0.9	0.509	0.509	0.043	0.407	0.642	0.590	0.723
Additive	0.0	0.329	0.324	0.061	0.209	0.489	0.000	0.297
	0.3	0.336	0.333	0.055	0.198	0.475	0.000	0.266
	0.6	0.366	0.371	0.047	0.243	0.500	0.010	0.286
	0.9	0.454	0.449	0.055	0.326	0.598	0.210	0.319

Table 3 Proportion of extreme PPP-values for the MH statistic and the MBC and PPP-value for the GDDM when a multidimensional model is used to simulate data and the unidimensional model to analyze data

Model	ρ	Proportion of extreme PPP-values		GDDM
		MH	MBC	PPP-value
Multi-uni	0.0	0.829	0.010	0.000
	0.3	0.762	0.610	0.000
	0.6	0.543	0.514	0.000
	0.9	0.057	0.029	0.000
Additive	0.0	0.867	0.838	0.000
	0.3	0.743	0.695	0.000
	0.6	0.390	0.248	0.000
	0.9	0.000	0.000	0.048

with exceptions of the two strongly correlated cases. The values of the GDDM-H again show good fit for all the cases (GDDM-H < 0.5), with the exception of the multi-unidimensional model with $\rho = 0.9$.

The results of the multidimensional data analyzed with a unidimensional model are reported in Tables 3 and 4. Here, we expect bad fit. From Table 3, we can notice that the proportion of extreme PPP-values for both the discrepancy measures decreases as the correlation among the trait increases. This outcome is coherent with the previous interpretation of nearly unidimensional data when the correlations among the latent variables are very strong. However, there is a further peculiar case, also observed in Matteucci and Mignani (2021a). In fact, the MH statistic and the MBC perform differently for the case of uncorrelated traits with multi-unidimensional data as, while the MH statistic correctly reports model misfit based on a rather high proportion of extreme PPP-values (0.867), the MBC fails by reporting only 1% of extreme PPP-values. The behavior of the MBC may be attributed to the peculiarity of the generating model, which resembles two separate unidimensional models. In particular, we have

Table 4 MBC-H and GDDM-H when a multidimensional model is used to simulate data and the unidimensional model to analyze data

Model	ρ	MBC-H						GDDM-H
		Mean	Median	SD	Min	Max	Prop > 0.5	
Multi-uni	0.0	0.644	0.597	0.140	0.466	0.924	0.876	1.000
	0.3	0.832	0.846	0.146	0.589	1.000	1.000	1.000
	0.6	0.813	0.836	0.135	0.581	1.000	1.000	1.000
	0.9	0.611	0.585	0.079	0.452	0.833	0.981	0.998
Additive	0.0	0.899	0.934	0.103	0.546	1.000	1.000	1.000
	0.3	0.877	0.891	0.098	0.613	1.000	1.000	1.000
	0.6	0.770	0.753	0.120	0.472	0.996	0.990	1.000
	0.9	0.551	0.550	0.065	0.343	0.741	0.829	0.825

noticed that the item discrimination estimates for the unidimensional model are higher than one for the first subtest (items 1-5) and close to zero for the second and the third subtest (items 6-15). This means that the latent trait is only defined by the first subtest items, which are related to the same ability. The PPP-values for the GDMM are all extreme (< 0.05), thus correctly reporting misfit for all the conditions.

With respect to the H distance, Table 4 shows that the average MBC-H is higher than the threshold of 0.5 for all the simulation conditions, reporting bad fit. Moreover, we observe a decrease in the MBC-H as the correlation among the traits increases, with the exception of the peculiar case of the multi-unidimensional data with uncorrelated traits described above. The proportions of MBC-H values higher than 0.5 is always close to 1, again showing bad fit. Lastly, the values of the GDDM-H equal to one show the maximum possible distance between the predictive discrepancy and the realized one, with slightly lower values only for strongly correlated traits.

The results of this simulation study show that the H distance is effective in highlighting the presence of possible misfit and determining plausible thresholds for classifying the misfit levels. In fact, unlike the approach based on the PPP-values, it is possible to evaluate the magnitude of the difference between the predictive and the realized distribution of the discrepancy measure (MBC or GDDM).

5 Empirical Application

The empirical data come from a survey conducted by the Center of Advanced Studies in Tourism-University of Bologna to investigate the residents' evaluations of their personal well-being in the Romagna area (see Bernini et al., 2018). The data used in this study consist of the responses given by 784 residents to 16 items of the questionnaire for the following three domains: personal Life (5 items), leisure activities

Table 5 Proportion of extreme PPP-values for the MH statistic and the MBC and PPP-value for the GDDM for the tourism data

Model	Proportion of extreme PPP-values		GDDM
	MH	MBC	PPP-value
Uni	0.642	0.633	0.000
Multi-uni	0.467	0.392	0.000
Additive	0.333	0.217	0.000
Bifactor	0.342	0.242	0.000

Table 6 DIC, WAIC, MBC-H, and GDDM-H for the tourism data

Model	DIC	WAIC	MBC-H						GDDM-H
			Mean	Median	SD	Min	Max	Prop > 0.5	
Uni	14038.79	13928.28	0.796	0.875	0.215	0.320	1.000	0.833	1.000
Multi-uni	12846.50	12621.08	0.629	0.610	0.238	0.219	1.000	0.625	1.000
Additive	12328.51	12083.12	0.505	0.466	0.238	0.097	1.000	0.475	1.000
Bifactor	12307.68	12061.63	0.497	0.450	0.247	0.084	1.000	0.442	1.000

(7 items), and life evaluation (4 items). Item responses have been dichotomized, where $Y = 1$ denotes satisfaction while $Y = 0$ indicates dissatisfaction.

The unidimensional, the multi-unidimensional, the additive, and the bifactor models are fitted via the Gibbs sampler. The bifactor model is a particular case of the additive model, where the overall and the specific latent variables are set orthogonal, while the correlations among the specific traits are estimated (see Gibbons & Hedeker, 1992). Three specific factors are assumed ($m = 3$), one for each domain. The same prior distributions used in the simulation have been used in the empirical application.

Table 5 reports the proportion of extreme PPP-values for the MH statistic and the MBC and the PPP-value estimate for the GDDM under the four competing models. It is clear that the unidimensional model is associated to the highest proportion of extreme PPP-values and that the fit improves as the model complexity increases. In particular, the additive and the bifactor models show similar performances. On the other hand, the PPP-value estimates for the GDDM are all equal to zero, reporting bad fit for all models.

Table 6 reports the results for the H distance, together with the deviance information criterion (DIC) and the Watanabe-Akaike information criterion (WAIC); see Gelman et al. (2014), Chap. 7. Both DIC and WAIC suggest that the additive and the bifactor models fit the data comparatively better than the unidimensional and the multi-unidimensional models, with a slight preference toward the bifactor one.

The results on the H distance are easily interpretable and confirm that the best models in terms of fit are the additive and the bifactor. In particular, the bifactor model is associated to an average MBC-H lower than 0.5 meaning that the amount

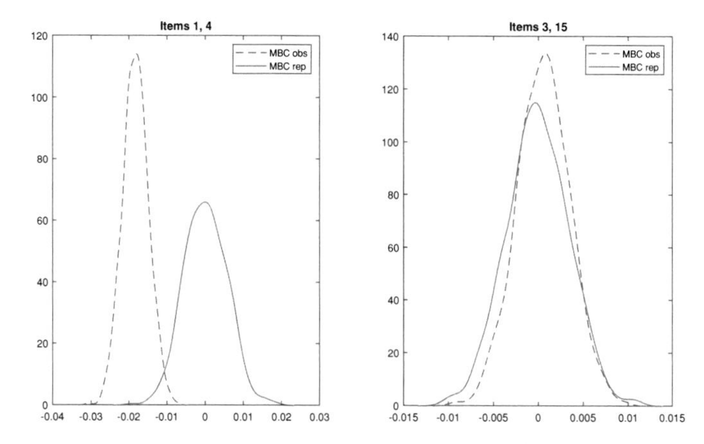

Fig. 1 Kernel densities of the realized (MBC obs) and predictive (MBC rep) discrepancies of MBC for the item pair 1, 4 (on the left) and the item pair 3, 15 (on the right)

of discrepancy between the predictive and the realized distribution of the MBC is the lowest among the competing models. This evidence is again confirmed with the proportion of item pairs for which the MBC-H is higher than 0.5 which is equal to 0.442 for the bifactor model, the lowest observed value. On the other hand, the GDDM-H reports bad fit for all the models and it is not able to compare the models effectively.

A strong advantage of the approach based on MBC-H is the possibility to investigate the fit of item pairs. In Fig. 1, we compare the realized (MBC obs) and the predictive distribution (MBC rep) of the MBC in the bifactor model for items 1 and 4 (on the left) and items 3 and 15 (on the right). The values of the H distance for the specific item pairs are $\text{MBC-H}_{1,4} = 0.9252$ and $\text{MBC-H}_{3,15} = 0.113$, indicating bad and good fit, respectively.

6 Concluding Remarks

The under-fitting scenario considered in this paper is particularly meaningful for its implications in real situations. In fact, unidimensional models are often used in practice even if unidimensionality and the following assumption of local independence do not hold. This often happens in behavioral or social sciences contexts, where more than one single domain affects the item responses.

The main strengths of the H distance, compared to traditional approaches, rely on the possibility (a) to directly quantify the amount of misfit, (b) to be used for model comparison purposes, and (c) to make more informative analyses on item pairs. Furthermore, it is demonstrated that, in practical applications, the MBC-H can be used to (a) leave out the models that show serious misfit by using the threshold of 0.5; (b) compare the amount of misfit of different competing models and choose the model which fits the data best; (c) identify, also through graphical plots, critical items that may involve misfit which are associated to high MBC-H in several pairs.

Future research may explore the use of different discrepancy measures that are able to investigate different aspects of model fit, other than the local independence assumption.

References

Béguin, A. A., & Glas, C. A. W. (2001). MCMC estimation and some fit analysis of multidimensional IRT models. *Psychometrika, 66*(4), 541–561. https://doi.org/10.1007/BF02296195

Bernini, C., Matteucci, M., & Mignani, S. (2018). Modelling subjective well-being dimensions through an IRT bifactor model: Evidences from an Italian study. *Electronic Journal of Applied Statistical Analysis, 11*(2), 427–446. https://doi.org/10.1285/i20705948v11n2p427.

Gelman, A., Carlin, J. B., Stern, H. S., Dunson, D. B., Vehtari, A., & Rubin, D. B. (2014). *Bayesian data analysis* (3rd ed.). CRC Press.

Gibbons, R. D., & Hedeker, D. R. (1992). Full-information item bi-factor analysis. *Psychometrika, 57*, 423–436. https://doi.org/10.1007/BF02295430

Levy, R., Mislevy, R. J., & Sinharay, S. (2009). Posterior predictive model checking for multidimensionality in item response theory. *Applied Psychological Measurement, 33*(7), 519–537. https://doi.org/10.1177/0146621608329504.

Levy, R., & Svetina, D. (2011). A generalized dimensionality discrepancy measure for dimensionality assessment in multidimensional item response theory. *British Journal of Mathematical and Statistical Psychology, 65*, 208–232. https://doi.org/10.1348/000711010X500483.

Matteucci, M., & Mignani, S. (2021). The Hellinger distance within posterior predictive assessment for investigating multidimensionality in IRT models. *Multivariate Behavioral Research, 56*(4), 627–648. https://doi.org/10.1080/00273171.2020.1753497.

Matteucci, M., & Mignani, S. (2021). Investigating model fit in item response models with the Hellinger distance. In Porzio, G. C., Rampichini, C., Bocci, C. (Eds.) *CLADAG 2021 book of abstracts and short papers* (pp. 150–153). Firenze University Press. https://doi.org/10.36253/978-88-5518-340-6.

Reckase, M. (2009). *Multidimensional item response theory*. Springer.

Rubin, D. B. (1984). Bayesianly justifiable and relevant frequency calculations for the applies statistician. *Annals of Statistics, 12*, 1151–1172. https://doi.org/10.2307/2240995.

Sheng, Y. (2008a). Markov chain Monte Carlo estimation of normal ogive IRT models in MATLAB. *Journal of Statistical Software*, *25*(8), 1–15. https://doi.org/10.18637/jss.v025.i08.

Sheng, Y. (2008b). A MATLAB package for Markov chain Monte Carlo with a multi-unidimensional IRT model. *Journal of Statistical Software*, *28*(10), 1–20. https://doi.org/10.18637/jss.v028.i10.

Sheng, Y. (2010). Bayesian estimation of MIRT models with general and specific latent traits in MATLAB. *Journal of Statistical Software*, *34*(3), 1–27. https://doi.org/10.18637/jss.v034.i03.

Sheng, Y., & Wikle, C. (2007). Comparing multiunidimensional and unidimensional item response theory models. *Educational and Psychological Measurement, 67*(6), 899–919. https://doi.org/10.1177/0013164406296977.

Sheng, Y., & Wikle, C. (2009). Bayesian IRT models incorporating general and specific abilities. *Behaviormetrika, 36*(1), 27–48. https://doi.org/10.2333/bhmk.36.27

Sinharay, S., Johnson, M. S., & Stern, H. S. (2006). Posterior predictive assessment of item response theory models. *Applied Psychological Measurement, 30*(4), 298–321. https://doi.org/10.1177/0146621605285517.

van der Linden, W. J., & Hambleton, R. K. (1997). *Handbook of modern item response theory*. Springer.

Wu, H., Yuen, K. V., & Leung, S. O. (2014). A novel relative entropy-posterior predictive model checking approach with limited information statistics for latent trait models in sparse 2^k contingency tables. *Computational Statistics & Data Analysis, 79,* 261–276. https://doi.org/10.1016/j.csda.2014.06.004.

Shapley-Lorenz Values for Credit Risk Management

Niklas Bussmann, Roman Enzmann, Paolo Giudici, and Emanuela Raffinetti

Abstract A trustworthy application of Artificial Intelligence requires to measure in advance its possible risks. When applied to regulated industries, such as banking, finance and insurance, Artificial Intelligence methods lack explainability and, therefore, authorities aimed at monitoring risks may not validate them. To solve this issue, explainable machine learning methods have been introduced to "interpret" black-box models. Among them, Shapley values are becoming popular: they are model agnostic and easy to interpret. However, they are not normalised and, therefore, cannot become a standard procedure for Artificial Intelligence risk management. This paper proposes an alternative explainable machine learning method, based on Lorenz Zonoids, that is statistically normalised, and can therefore be used as a standard for the application of Artificial Intelligence. The empirical analysis of 15,000 small and medium companies asking for credit confirms the advantages of our proposed method.

Keywords Artificial intelligence · Lorenz Zonoids · Shapley values · Risk management

N. Bussmann · P. Giudici · E. Raffinetti (✉)
Department of Economics and Management, University of Pavia, Via San Felice al Monastero 5, 27100 Pavia, Italy
e-mail: emanuela.raffinetti@unipv.it

N. Bussmann
e-mail: niklas.bussmann01@universitadipavia.it

P. Giudici
e-mail: paolo.giudici@unipv.it

R. Enzmann
University of Bonn, Bonn, Germany
e-mail: ryenzmann@hotmail.com

L. Grilli et al. (eds.), *Statistical Models and Methods for Data Science*, Studies in Classification, Data Analysis, and Knowledge Organization,
https://doi.org/10.1007/978-3-031-30164-3_10

1 Introduction

A key prerequisite to develop reliable and trustworthy Artificial Intelligence (AI) methods is the capability to measure their risks. When applied to high impact and regulated industries, such as energy, finance and health, AI methods need to be validated by national regulators, possibly according to international standards, such as ISO/IEC CD 23894. Most AI methods rely on the application of highly complex machine learning models which, while reaching high predictive performance, lack explainability. This is a problem in regulated industries, as authorities aimed at monitoring the risks arising from the application of AI methods may not validate them (see, e.g., Financial Stability Board 2020).

In these fields, comprehensible results need to be obtained to allow organisations to detect risks, especially in terms of the factors which can cause them. This objective is more evident when dealing with AI systems.

Indeed, AI methods have an intrinsic black-box nature, resulting in automated decision-making that can, for example, classify a person into a class associated with the prediction of its individual behaviour, without explaining the underlying rationale. To avoid that wrong actions are taken as a consequence of "automatic" choices, AI methods need to explain the reasons for their classifications and predictions.

The notion of "explainable" AI has become very important in the recent years, following the increasing application of AI methods that impact the daily life of individuals and societies. At the institutional level, explanations can answer different kinds of questions about a model's operations depending on the stakeholder they are addressed to (see, e.g., Bracke et al. 2019): developers, managers, model checkers and regulators. In general, to be explainable, AI methods have to provide details or reasons clarifying their functioning.

From a mathematical viewpoint, the explainability requirement can be fulfilled using simple machine learning models (such as logistic and linear regression models). However, these models provide reduced predictive accuracy. To improve predictive accuracy, the implementation of complex machine learning models (such as neural networks and random forests) seems necessary, but this leads to limited interpretability.

This trade-off can be solved empowering accurate machine learning models with innovative methodologies able to explain their classification and prediction output. A recent attempt in this direction can be found in Bussmann et al. (2020), who proposed to apply correlation networks (see, e.g., Mantegna and Stanley 1999) to Shapley values (see Shapley 1953) so that AI predictions can be grouped according to the similarity in the underlying explanations. The proposal has been applied to the scoring of borrowers in financial lending, a service for which the use of AI methods is developing fast.

Shapley values have the advantage of being agnostic: independent on the underlying model with which classifications and predictions are computed; but have the disadvantage of not being normalised and, therefore, difficult to be used in comparisons outside the specific application.

In this paper, we propose an alternative explainable machine learning methods, based on the combination between the Shapley value approach (see Shapley 1953) and the Lorenz Zonoid tool described in Giudici and Raffinetti (2020). Shapley values belong to the class of local explanation methods, as they aim to interpret individual predictions in terms of which variables mostly affect them (see, e.g., Financial Stability Board 2020 and Joseph 2019). Lorenz Zonoids instead are a global explanation method, as they aim to interpret all model predictions as a whole, in terms of which variables mostly determine them, for all observations. The combination between the two approaches has been successfully tested in Giudici and Raffinetti (2021) within the context of predicting a continuous variable, the price of bitcoins, by means of three well-known continuous financial variables: the price of oil, gold and the Standard and Poor's index.

In this paper, we extend the methodology to a much more challenging problem: the prediction of credit default (binary variable) by means of a large set of highly correlated company performance variables, taken from balance sheets. This requires extending the statistical part of the methodology, to binary response variables and binary classification, and the computational part of the proposal, providing an automated search through a high-dimensional space of possible alternative models.

Our empirical findings are very promising and may lead to a new explainable machine learning method to measure financial risk management, based on the Shapley-Lorenz approach which, different from Shapley values, gives rise to normalised values that can be used in any comparisons between alternative model explanations.

The paper is organised as follows: the next section illustrates the methodology; Sect. 3 provides the Python algorithm that can be employed to apply our proposed method; Sect. 4 discusses the empirical findings obtained by applying our proposal to financial data; finally, Sect. 5 contains some brief concluding remarks.

2 Methodology

To meet the requirement that risk measurement is explainable, leading to develop a trustworthy application of AI, in this section we propose a global explainable machine learning method. Our proposal derives from the combination of two research streams. The first one concerns the development of machine learning methods for binary classification problems. The second one concerns the development of explainable methods to understand the output of advanced machine learning models. The result is a novel method for AI risk management which is, at the same time, predictively accurate, interpretable and robust to anomalies in the observations.

2.1 Binary Classification

Let Y be a binary response variable, which can, for example and without loss of generality, express whether a company defaults ($Y = 1$) or not ($Y = 0$). Given K explanatory variables $X_1, \ldots, X_K$, a logistic regression model for Y can be specified as follows:

$$ln\left(\frac{\pi_i}{1-\pi_i}\right) = \beta_0 + \sum_{k=1}^{K} \beta_k x_{ik} = \eta_i, \tag{1}$$

where $i = 1, \ldots, n; \eta_i = \beta_0 + \sum_{k=1}^{K} \beta_k x_{ik}; \pi_i$ represents the probability of the event for the i-th observation (company); $\mathbf{x}_i = (x_{i1}, \ldots, x_{ik})$ points out the K-dimensional vector reporting the values taken by the K explanatory variables referring to the i-th observation; β_0 and β_k are the parameters representing the intercept and the k-th regression coefficient, respectively.

By means of the Maximum Likelihood Estimation method, the parameters β_0 and β_k can be estimated leading to derive the predicted probability of the default as

$$\hat{\pi}_i = \frac{e^{\eta_i}}{1 + e^{\eta_i}}, \tag{2}$$

which can be employed to attach to the i-th observation a "score", a number between zero and one which can be interpreted to signal, for example, the creditworthiness of a company: the higher the score, the lower the trust. A classification of each company can then follow, comparing the score with an appropriate threshold, chosen on an experiential basis.

While in the following we will refer to the specific case of credit scoring, what was proposed is quite general, and can be applied to any binary classification problems.

2.2 The Shapley-Lorenz Decomposition for Credit Risk Data

Following the early experimentation in Giudici and Raffinetti (2021), we propose, for financial risk management purposes, a global explainable AI method, named Shapley-Lorenz decomposition, which combines the interpretability power of the local Shapley value game theoretic approach (see Shapley 1953) with a more robust global approach based on the Lorenz Zonoid model accuracy tool (see Giudici and Raffinetti 2020). The Lorenz Zonoids, originally introduced by Koshevoy and Mosler (1996), were further developed by Giudici and Raffinetti (2020) as a generalisation of the ROC curve in a multidimensional setting and, therefore, the Shapley-Lorenz decomposition has the advantage of combining predictive accuracy and explainability performance into one single diagnostics. Furthermore, the Lorenz Zonoid is based on a measure of mutual variability that is more robust to the presence of outlying (anomalous) observations with respect to the standard variability around the mean.

These theoretical properties can be exploited to develop partial dependence measures that allow to detect the additional contribution of a new predictor into an existing model.

Shapley values were originally introduced by Shapley (1953) as a pay-off concept from cooperative game theory. When referring to machine learning models, the notion of pay-off corresponds to the model prediction. For any single statistical unit i $(1 = 1, \ldots, n)$, the pay-offs are computed as

$$p_{off}(X_i^k) = \hat{f}(X^{'} \cup X_k)_i - \hat{f}(X^{'})_i, \tag{3}$$

where $\hat{f}(X^{'})_i$ are the predicted values provided by a machine learning model, depending only on $X^{'}$ predictors; $\hat{f}(X^{'} \cup X_k)_i$ are the predicted values generated by the machine learning model, depending both on the $X^{'}$ predictors and the additional included X_k predictor.

The main advantage of Shapley values, over alternative explainable AI (XAI) methods, is that they can be employed to measure the contribution of each explanatory variable for each point prediction of a machine learning model, regardless of the underlying model itself (see, e.g., Lundberg and Lee 2017, Štrumbelj and Kononenko 2010). In other words, Shapley-based XAI methods combine generality of application (they are model agnostic) with the personalisation of their results (they can explain any single point prediction).

The main drawback of Shapley values is that they provide explainability scores that are not normalised. They can be used to compare the relative contribution of one variable to that of another, but they cannot be used to assess the absolute importance of each variable, and to make comparisons beyond the specific context.

The key benefit related to the employment of the Lorenz Zonoid tool is the possibility of evaluating the contribution associated with any additional explanatory variable to the whole model prediction with a normalised measure that can be used to assess the importance of each variable.

The Lorenz Zonoid measure was introduced by Giudici and Raffinetti (2020) to develop new partial dependence measures. Specifically, given a set of K explanatory variables, the marginal contribution associated with the variable X_k can be determined into a global perspective by resorting to the difference between two Lorenz Zonoids. More precisely, the two Lorenz Zonoids are built on the predictions $\hat{Y}_{X' \cup X_k}$, provided by a model including the additional covariate X_k, and on the predictions $\hat{Y}_{X'}$, provided by the reduced model excluding the covariate X_k. The additional contribution (AC) related to the inclusion of covariate X_k can be then expressed as

$$AC = LZ(\hat{Y}_{X' \cup X_k}) - LZ(\hat{Y}_{X'}), \tag{4}$$

where $LZ(\hat{Y}_{X' \cup X_k})$ and $LZ(\hat{Y}_{X'})$ define the Lorenz Zonoids computed on the predicted values provided by the model including also covariate X_k and the Lorenz Zonoids computed on the predicted values provided by the model including the $X^{'}$ covariates but excluding covariate X_k, respectively.

As we are dealing with a binary response variable, denoting the active and default status of the companies, the terms $\hat{Y}_{X' \cup X_k}$ and $\hat{Y}_{X'}$ can be re-written as the predicted probabilities of default $\hat{\pi}_{X' \cup X_k}$ and $\hat{\pi}_{X'}$, when resorting to the logistic regression model including also the explanatory variable X_k and to the logistic regression model not including the explanatory variable X_k, respectively. Thus, equation in (4) becomes

$$AC = LZ(\hat{\pi}_{X' \cup X_k}) - LZ(\hat{\pi}_{X'}). \tag{5}$$

In order to compute the Lorenz Zonoids $LZ(\hat{\pi}_{X' \cup X_k})$ and $LZ(\hat{\pi}_{X'})$ in Eq. (5), we resort to the covariance operators (see, e.g., Lerman and Yitzhaki 1984), i.e.,

$$LZ(\hat{\pi}_{X' \cup X_k}) = \frac{2Cov(\hat{\pi}_{X' \cup X_k}, r(\hat{\pi}_{X' \cup X_k}))}{nE(\hat{\pi}_{X' \cup X_k})} \quad \text{and}$$

$$LZ(\hat{\pi}_{X'}) = LZ(\hat{\pi}_{X'}) = \frac{2Cov(\hat{\pi}_{X'}, r(\hat{\pi}_{X'}))}{nE(\hat{\pi}_{X'})}, \tag{6}$$

where $r(\hat{\pi}_{X' \cup X_k})$ and $r(\hat{\pi}_{X'})$ are the rank scores of $\hat{\pi}_{X' \cup X_k}$ and $\hat{\pi}_{X'}$; n is the sample size; $E(\hat{\pi}_{X' \cup X_k})$ and $E(\hat{\pi}_{X'})$ are the expected values of $\hat{\pi}_{X' \cup X_k}$ and $\hat{\pi}_{X'}$.

The Shapley-Lorenz decomposition expression is the result of a combination between the Shapley value-based formula and the Lorenz Zonoid tools. Formally, the contribution of the additional variable X_k, expressed in terms of the differential contribution to the global predictive accuracy, equals

$$LZ^{X_k}(\hat{\pi}) = \sum_{X' \subseteq \mathcal{C}(X) \setminus X_K} \frac{|X'|!(K - |X'| - 1)!}{K!} [LZ(\hat{\pi}_{X' \cup X_k}) - LZ(\hat{\pi}_{X'})], \tag{7}$$

where $LZ(\hat{\pi}_{X' \cup X_k})$ and $LZ(\hat{\pi}_{X'})$ measure the marginal contribution provided by the inclusion of variable X_k; K is the number of available predictors; $\mathcal{C}(X) \setminus X_k$ is the set of all the possible model configurations which can be obtained with $K - 1$ variables, excluding variable X_k; $|X'|$ denotes the number of variables included in each possible model.

Finally, it is worth noting that the Shapley-Lorenz decomposition presents as an agnostic explainable Artificial Intelligence method which can be applied to the predictive output, regardless of which model and data generated it.

3 Algorithm

The Python code that we used to compute the Shapley-Lorenz Zonoid values is available on https://github.com/roye10/ShapleyLorenz, for full reproducibility. We now summarise the logical structure of the code.

Table 1 Shapley-Lorenz Zonoid algorithm

0.	A pre-defined (user supplied) upper bound on total number of fully considered subset permutations can be defined, given, by say, *n_perm*. This is similar to the *nsamples* parameter used in the kernel SHAP module by Lundberg and Lee (2017). Unlike the kernel SHAP module, however, only subsets are considered, for which full permutations can be considered, given *n_perm*. Subsets are considered sequentially, in order of highest to lowest Shapley kernel weights, defined by $\frac{\lvert X'\rvert!(K-\lvert X'\rvert-1)!}{K!}$. Due to the symmetric property of this Shapley kernel weight, a given subset is always considered pair wise, with its complement, i.e. first all permutations of subset size 1 and those of size $K-1$ are considered. If all permutations of the next subset can be considered as well, given the upper bound in *n_perm*, this is added to the subset sizes considered in the next step.
1.	***Do for*** $k \in K$: (i.e. for all covariates)
1(a)	***Do for*** $s \in \mathcal{C}(X)$ (i.e. for all subset permutations):
1(b)	***Do for*** $i \in N$: Let $\tilde{X}$ contain a given permutation of a given subset size without k. Compute $E[f(X) \mid \tilde{X} = \tilde{X}_i]$. Once for $\tilde{X}/k$ and once for $\tilde{X} \cup X_k$. Assuming covariate independence, this can be approximated by $\frac{1}{N}\sum_{j=1}^{n} f(X_j/\tilde{X}, \tilde{X}_i)$, and analogously for $\tilde{X}_i \cup X_k$. $X/\tilde{X}$ represents all covariates not included in the subset $X' \cup X_k$ and is obtained by replacing those covariates with either training data or a row-wise shuffled variant of the original covariate matrix. The result of this step, thus is $E[f(X) \mid \tilde{X} = \tilde{X}_i]$, approximated by the sample mean, over the underlying distribution of X.
2.	Sort the obtained values for the current permutation of the current subset size iteration.
3.	Compute the Lorenz Zonoid share for the current permutation of the current subset size iteration. Once for the permutation not including k and once for the permutation including k. Then compute the difference.
4.	Weight obtained Lorenz Zonoid difference, by the kernel weight as defined above.
5.	Compute weighted sum of differences.

Following the Shapley value attribution method, computing exact Shapley-Lorenz Zonoid covariate contribution measures for K covariates requires the computation of Lorenz Zonoid marginal contributions across 2^K different subsets, per covariate. Computationally, this becomes intractable for non-conservative covariate sizes, and therefore an approximate solution is implemented in the Shapley-Lorenz Zonoid package. The method can be summarised as in Table 1.

We remark that as the Shapley-Lorenz Zonoid approach is based on the construction of all the possible model configurations, it results computationally intensively. To overcome this drawback, especially when the set of the available explanatory variables is quite large, a preliminary selection of the predictors to be included into the model may be led through the employment of the Lorenz Zonoid-based measures proposed in Giudici and Raffinetti (2020).

4 Application

4.1 Data

We apply our proposed method to data supplied by European External Credit Assessment Institution (ECAI) that specialises in credit scoring for P2P platforms focused on SME commercial lending. The data are described by Giudici et al. (2020) to which we refer for further details. In summary, the analysis relies on a dataset composed of official financial information, extracted from the balance-sheets of 15,045 SMEs, mostly based in Southern Europe, for the year 2015. The information about the status (0 = active, 1 = defaulted) of each company one year later (2016) is also provided. The observed proportion of defaulted companies is equal to 10.9%.

The nineteen financial variables included in our dataset are as follows: Total Assets/Equity; (Long term debt+Loans)/Shareholders Funds; Total Assets/Total Liabilities; Current Assets/Current Liabilities; (Current assets-Current assets: stocks)/ Current liabilities; (Shareholders Funds+Non current liabilities)/Fixed assets; EBIT/ interest paid; (Profit or Loss before tax+Interest paid)/Total assets; Return on Equity; Operating revenues/Total assets; Sales/Total assets (Activity Ratio); Interest paid/(Profit before taxes+Interest paid); EBITDA/interest paid; EBITDA/Operating revenues; EBITDA/Sales; Trade Payables/Operating Revenues; Trade Receivables/ Operating Revenues; Inventories/Operating Revenues; Turnover. We remark that, for these variables, and particularly for those reflecting the operations of the companies, there is a noticeable presence of unusually large or small values when compared to the mean. These outliers should not be substituted or deleted as they can provide important explanations on the companies included in the sample. However, their presence is going to affect the robustness of machine learning models, and of Shapley values in particular. This further motivates the use of Shapley-Lorenz values, more robust to outliers.

4.2 Results

Using the data, Giudici et al. 2020 have constructed logistic regression scoring models that aim at estimating the probability of the default of each company, using the available financial data from the balance sheets and, in addition, network centrality measures that are obtained from similarity networks.

To improve the predictive performance of the model, Bussmann et al. (2020) have applied to the same database the Gradient Boosting Tree algorithm, and obtained a substantial increase in predictive performance.

The same authors identify Variable 3: Total Assets/Total Liabilities; Variable 7: EBITDA/Interest paid; Variable 8: (Profit or Loss before tax + Interest paid)/Total asset as the variables that mostly contribute to Shapley value decomposition. This is quite consistent with most credit scoring models that typically include, among the explanatory variables of credit default, a measure of financial leverage (such as Variable 3) and a measure of profitability (such as Variables 7 and 8).

We use the same data and apply the proposed Shapley-Lorenz approach procedure to detect the main factors impacting the probability of default. Although the Gradient Boosting Tree method is more performing in terms of predictive accuracy, providing an Area Under the ROC Curve (AUROC) equal to 0.93, we resort to the logistic regression model to fit our data. This is because, if on the one hand the logistic regression model implementation leads to a lack in terms of predictive accuracy (as the AUROC decreases to a value of 0.81), on the other hand, it appears as a model explainable by design and then results simpler to be understood.

Through the cross-validation procedure, the data is split into a training set (80%) and a test set (20%). We then calculate, on the same split, the contribution of each of the nineteen explanatory variables to the estimate of the probability of default, using two explainable AI methods: the Shapley value approach and the Lorenz-Shapley approach that we propose.

Table 2 contains the results of the comparison. For each variable, we report the value of the Shapley-Lorenz Zonoid contribution and the Shapley Value contribution, calculated as the sum of the Shapley values over all observations. For comparison purposes, we also report the contribution of each variable to the deviance (G^2), calculated using the Shapley value formula, as in the Global ANOVA—like explainable AI method described in Financial Stability Board (2020) and Joseph (2019).

From Table 2, note that the variables which most contribute to the prediction of default, according to the sum of the Shapley values, are Variable 8: (Profit or Loss before tax + Interest paid)/Total asset; followed at a considerable distance by Variable 13 and 14 (both related to EBITDA) and by Variable 3 (Total Assets/Total liabilities). In terms of G^2, instead, the differences between Variable 8 (the highest contributor) and Variables 14, 15 and 3 are lower. The role that Variable 13 has in terms of Shapley value been replaced by Variable 15. Note that the variables selected as the most important by both Shapley values and G^2 have a similar structure: one variable that indicates leverage (Variable 3) and a few variables that express profitability (Variables 8, 13, 14 or Variables 8, 14, 15). The latter are highly correlated, as they are based on similar information: this may indicate a weakness of the selection methods, possibly redundant and more sensible to outlier observations.

The first column of Table 2, giving the Shapley-Lorenz values, indicate, instead, that Variable 8, with a value of 0.165, and Variable 3, with a value of 0.114, are one magnitude order higher than the others. This indicates a more clearcut choice,

Table 2 Marginal contribution of each explanatory variable in terms of Shapley-Lorenz Zonoids, G^2 and total Shapley values

ID	Variable	Shapley-Lorenz	G^2	Shapley
1	Total assets/Equity	0.003	0.16	2.53
2	(Long term debt + Loans)/Shareholders Funds	0.002	0.54	–202.80
3	Total assets/Total liabilities	0.114	1088.12	–1273.97
4	Current assets/Current liabilities	0.057	553.68	–641.69
5	(Current assets – Current assets: stocks)/Current liabilities	0.001	479.06	–93.51
6	(Shareholders Funds + Non current liabilities)/Fixed assets	0.003	13.16	4180.56
7	EBIT/interest paid	–0.001	411.10	1504.44
8	(Profit or Loss before tax + Interest paid)/Total assets	0.165	1633.51	–13115.53
9	Return on Equity	0.054	826.96	–1993.98
10	Operating revenues/Total assets	0.062	17.36	–289.46
11	Sales/Total assets	–0.002	10.96	252.59
12	Interest paid/(Profit before taxes + Interest paid)	0.013	103.26	379.73
13	EBITDA/interest paid	0.024	418.00	–1697.31
14	EBITDA/Operating revenues	0.038	1254.63	–1419.43
15	EBITDA/Sales	0.026	1122.05	–785.95
16	Trade Payables/Operating revenues	0.001	14.73	–193.60
17	Trade Receivables/Operating revenues	0.058	475.40	–585.58
18	Inventories/Operating revenues	0.013	126.78	1190.47
19	Turnover	0.022	85.26	1072.37

with only two variables being selected: a measure of leverage, and a measure of profitability. In the latter case, only the most contributing one, among the several that measure profitability, is chosen.

We remark that the negative sign appearing for Variables 7 and 11 is not to be intended as the presence of an indirect relationship with the probability of default, as in the case of Shapley values. Indeed, the Shapley-Lorenz values being normalised, the negative sign is due to the approximation internal to the algorithm.

Comparing the results obtained with the different methods, the Shapley-Lorenz selection is evidently the easiest to interpret and, by construction, it is more robust to outlier observations, as can be easily noticed by repeating the analysis for different training/validation splits.

Further research focuses on the application of the methodology to other Artificial Intelligence applications that involve a binary classification problem. Our code is openly accessible for this purpose. This may allow the development of a quantitative standard for AI risk management.

5 Concluding Remarks

The paper proposes a new agnostic explainable AI method, based on the combination of Shapley values and Lorenz Zonoids, that can be used to interpret the results of a highly performing machine learning risk management algorithm.

The proposed method, like other explainable AI methods, is able to identify the variables which most affect the predictions, combining the notion of explainability with that of predictive accuracy.

The application of the proposal to a credit risk management use case shows its superior performance, in terms of selectivity, consistency with economic intuition and robustness. It identifies one measure of leverage and one measure of profitability as those that most matter.

We can thus conclude that the illustrated method is satisfactory and can be proposed as a use case to standardise risk measurement and management in the application of Artificial Intelligence to credit lending.

The implementation of the Shapley-Lorenz approach to other application fields with further linear and non-linear models represents an interesting research area which is currently in progress. Moreover, due to its appealing features, the Shapley-Lorenz decomposition can be further extended to fulfil the requirements of robustness and fairness which, together with the explainability and accuracy principles, contribute to ensuring a trustworthy AI. Indeed, as the Lorenz Zonoid appears as a measure of mutual variability, it can be exploited to both provide robust findings (i.e., not affected by outlying observations) and measures of fairness (equality) with respect to specific sub-groups of the population.

References

Bracke, P., Datta, A., Jung, C., & Shayak, S. (2019). Machine learning explainability in finance: An application to default risk analysis. https://www.bankofengland.co.uk/working-paper/2019/machine-learning-explainability-in-finance-an-application-to-default-risk-analysis.

Bussmann, N., Giudici, P., Marinelli, D., & Papenbrock, J. (2020). Explainable machine learning in credit risk management. *Computational Economics, 57*, 203–216. https://doi.org/10.1007/s10614-020-10042-0

Financial Stability Board: Interpretable Machine Learning-A Guide for Making Black Box Models explainable (2020). https://cristophm.github.io/interpretable-ml-book.

Giudici, P., & Raffinetti, E. (2020). Lorenz model selection. *Journal of Classification, 37*(2), 754–768. https://doi.org/10.1007/s00357-019-09358-w

Giudici, P., & Raffinetti, E. (2021). Shapley-Lorenz explainable artificial intelligence. *Expert Systems with Applications, 167*, 1–9. https://doi.org/10.1016/j.eswa.2020.114104

Giudici, P., Hadji-Misheva, B., & Spelta, A. (2020). Network based credit risk models. *Quality Engineering, 32*(1), 199–211. https://doi.org/10.1080/08982112.2019.1655159

Joseph, A. (2019). Parametric inference with universal function approximators. https://www.bankofengland.co.uk/working-paper/2019/shapley-regressions-a-framework-for-statistical-inference-on-machine-learning-models.

Koshevoy, G., & Mosler, K. (1996). The Lorenz Zonoid of a multivariate distribution. *Journal of the American Statistical Association, 91*(434), 873–882. 10/2291682.

Lerman, R., & Yitzhaki, S. (1984). A note on the calculation and interpretation of the Gini index. *Economics Letters, 15*(3–4), 363–368. https://doi.org/10.1016/0165-1765(84)90126-5

Lundberg, S. M., & Lee, S. (2017). A unified approach to interpreting model predictions. In *Conference on Neural Information Processing Systems (NIPS 2017)*. Long Beach.

Mantegna, R. N., & Stanley, H. E. (1999). *Introduction to econophysics: Correlations and Complexity in finance*. Cambridge University Press.

Shapley, L. S. (1953). A value for *n*-person games. In H. Kuhn & A. Tucker (Eds.), *Contributions to the theory of games II* (pp. 307–317). Princeton University Press.

Štrumbelj, E., & Kononenko, I. (2010). An efficient explanation of individual classifications using game theory. *Journal of Machine Learning Research, 11*, 1–18.

A Study of Lack-of-Fit Diagnostics for Models Fit to Cross-Classified Binary Variables

Maduranga Dassanayake and Mark Reiser

Abstract In this paper, an extended version of the *GFfit* statistic is compared to other lack-of-fit diagnostics for models fit to cross-classified binary variables. The extended *GFfit* statistic is obtained by decomposing the Pearson statistic from the full table into orthogonal components defined on marginal distributions. The extended version of the statistic, $GFfit_{\perp}^{(ij)}$, can be applied to a variety of models for cross-classified tables. Simulation results show that $GFfit_{\perp}^{(ij)}$ has good Type I error performance even if the joint frequencies are very sparse. Asymptotic power calculations and simulation results show that $GFfit_{\perp}^{(ij)}$ has higher power for detecting the source of lack of fit compared to other diagnostics on bivariate marginal tables for binary variables.

Keywords Multinomial distribution · Pearson Chi-square · Orthogonal components · IRT model

1 Introduction

For a multi-way contingency table, the traditional Pearson's Chi-square statistic is obtained by comparing the observed frequencies to the expected frequencies under the null hypothesis. A composite null hypothesis $H_0\colon \boldsymbol{\pi} = \boldsymbol{\pi}(\boldsymbol{\beta})$, where the null distribution depends on a vector of g unknown parameters $\boldsymbol{\beta} = (\beta_1, \ldots, \beta_g)'$, can be tested with the Pearson-Fisher statistic, $X^2_{PF} = \sum_s z_s^2$, where $z_s = \sqrt{n}(\pi_s(\hat{\boldsymbol{\beta}}))^{-\frac{1}{2}}$ $\left(\hat{\mathrm{p}}_s - \pi_s(\hat{\boldsymbol{\beta}})\right)$, and where $\hat{\mathrm{p}}_s$ is element s of $\hat{\mathbf{p}}$, a vector of multinomial proportions, n is the total sample size, $\hat{\boldsymbol{\beta}}$ is the parameter estimator vector, $\pi_s(\boldsymbol{\beta})$ is the expected

M. Dassanayake
Department of Statistics, University of Georgia, Athens, GA, USA
e-mail: maduranga@uga.edu

M. Reiser (✉)
School of Mathematical and Statistical Sciences, Arizona State University, Tempe, AZ, USA
e-mail: mark.reiser@asu.edu

L. Grilli et al. (eds.), *Statistical Models and Methods for Data Science*, Studies in Classification, Data Analysis, and Knowledge Organization,
https://doi.org/10.1007/978-3-031-30164-3_11

proportion for cell s and $\pi_s(\hat{\boldsymbol{\beta}})$ is the estimated expected proportion for cell s. The degrees of freedom for a model on q cross-classified variables, $2^q - g - 1$, were given in Fisher (1924). Orthogonal components of X^2_{PF} have been studied by many authors, including (Lancaster 1969) and (Rayner and Best 1989). A new extended *GFfit* statistic was proposed in Reiser et al. (2022) for the purpose of detecting lack of fit. The extended statistic, $GFfit_{\perp}^{(ij)}$, is obtained by decomposing the Pearson statistic from the full table into orthogonal components defined on marginal distributions. In this paper, the performance of $GFfit_{\perp}^{(ij)}$ is compared to standardized residuals (Reiser 1996) and $\bar{\chi}^2_{ij}$ (Liu and Maydeu-Olivares 2014) using simulations to assess Type I error rate and power for a latent variable model fit to binary cross-classified variables. $\bar{\chi}^2_{ij}$ is a moment corrected version of χ^2_{ij}, where χ^2_{ij} is the Pearson statistic applied to bivariate table i, j.

2 Marginal Proportions

A traditional statistic such as X^2_{PF} uses the joint frequencies to calculate goodness of fit for a model that has been fit to a cross-classified table. This section presents a transformation from joint proportions or frequencies to marginal proportions. The method for using marginal proportions to obtain the $GFfit_{\perp}^{(ij)}$ and other statistics mentioned in Sect. 1 are explained in Sect. 3.

2.1 First- and Second-Order Marginals

The relationship between joint proportions and marginals can be shown by using 0's and 1's to code the levels of dichotomous response random variables, $Y_i, i = 1, 2, \ldots, q$, where Y_i follow the Bernoulli distribution with parameter P_i. Then, a q-dimensional vector of 0's and 1's, sometimes called a response pattern, will indicate a specific cell from the contingency table formed by the cross-classification of q response variables. For dichotomous response variables, a response pattern is a sequence of 0's and 1's with length q. The $T = 2^q$-dimensional set of response patterns can be generated by varying the levels of the qth variable most rapidly, the $q^{th} - 1$ variable next, etc. Define $\boldsymbol{V}$ as the T by q matrix with response patterns as rows. For instance when $q = 3$,

$$V = \begin{pmatrix} 0\,0\,0 \\ 0\,0\,1 \\ 0\,1\,0 \\ 0\,1\,1 \\ 1\,0\,0 \\ 1\,0\,1 \\ 1\,1\,0 \\ 1\,1\,1 \end{pmatrix}.$$

Let v_{is} represent element i of response pattern s, $s = 1, 2, \ldots, T$. Then, under the model $\boldsymbol{\pi} = \boldsymbol{\pi}(\boldsymbol{\beta})$, the first-order marginal proportion for variable Y_i can be defined as $P_i(\boldsymbol{\beta}) = \text{Prob}(Y_i = 1|\boldsymbol{\beta}) = \sum_s v_{is}\pi_s(\boldsymbol{\beta})$, and the true first-order marginal proportion is given by $P_i = \text{Prob}(Y_i = 1) = \sum_s v_{is}\pi_s$. Under the model, the second-order marginal proportion for variables Y_i and Y_j can be defined as $P_{ij}(\boldsymbol{\beta}) = \text{Prob}(Y_i = 1, Y_j = 1|\boldsymbol{\beta}) = \sum_s v_{is}v_{js}\pi_s(\boldsymbol{\beta})$, where $j = 1, 2, \ldots, q-1$; $i = j+1, \ldots q$, and the true second-order marginal proportion is given by $P_{ij} = \text{Prob}(Y_i = 1, Y_j = 1) = \sum_s v_{is}v_{js}\pi_s$.

2.2 *Higher Order Marginals*

A general matrix $\mathbf{H}_{[t:u]}$ to obtain marginals of any order can be defined by using Hadamard products among the columns of $\boldsymbol{V}$. The symbol $\mathbf{H}_{[t:u]}$, $t \leq u \leq q$ denotes the transformation matrix that would produce marginals from order t up to and including order u. Furthermore, $\mathbf{H}_{[t]} \equiv \mathbf{H}_{[t:t]}$. $\mathbf{H}_{[1:q]}$ gives a mapping from joint proportions to the set of $(2^q - 1)$ marginal proportions: $\boldsymbol{P} = \mathbf{H}_{[1:q]}\boldsymbol{\pi}$, where $\boldsymbol{P} = (P_1,\ P_2,\ P_3, \ldots P_q,\ P_{12},\ P_{13}, \ldots P_{q-1,q},\ P_{1,1,2} \ldots P_{q-2,q-1,q} \ldots P_{1,2,3 \ldots q})'$ is the vector of marginal proportions. An example of $\mathbf{H}_{[1:3]}$ for $q = 3$ is given in the online supplement.

3 Lack-of-Fit Statistics

This section illustrates the mathematical details of the $GFfit_{\perp}^{(ij)}$ statistic, adjusted residuals and the $\bar{\chi}_{ij}^2$ statistic mentioned in Sect. 1.

3.1 The $GFfit_{\perp}^{(ij)}$ Statistic

A traditional composite null hypothesis for a test of fit on a multinomial model is $H_0: \boldsymbol{\pi} = \boldsymbol{\pi}(\boldsymbol{\beta})$. Linear combinations of $\boldsymbol{\pi}$ may be tested under the null hypothesis $H_0: \mathbf{H}\boldsymbol{\pi} = \mathbf{H}\boldsymbol{\pi}(\boldsymbol{\beta})$. $\mathbf{H}$ may specify linear combinations that form marginal proportions as defined in Sect. 2.2.

Define the unstandardized residual $\mathrm{u}_s = \hat{\mathrm{p}}_s - \pi_s(\hat{\boldsymbol{\beta}})$, and denote the vector of unstandardized residuals as $\mathbf{u}$ with element u_s. $\sqrt{n}\ \mathbf{u}$ has asymptotic covariance matrix $\boldsymbol{\Omega}_{\mathbf{u}}$, where

$$\boldsymbol{\Omega}_{\mathbf{u}} = (D(\boldsymbol{\pi}(\boldsymbol{\beta})) - \boldsymbol{\pi}(\boldsymbol{\beta})\boldsymbol{\pi}(\boldsymbol{\beta})' - \mathbf{G}(\mathbf{A}'\mathbf{A})^{-1}\mathbf{G}'), \tag{3.1}$$

and where

$$D(\boldsymbol{\pi}(\boldsymbol{\beta})) = \text{diagonal matrix with } (s, s) \text{ element equal to } \pi_s(\boldsymbol{\beta}),$$

$$\mathbf{A} = D(\boldsymbol{\pi}(\boldsymbol{\beta}))^{-1/2}\frac{\partial \boldsymbol{\pi}(\boldsymbol{\beta})}{\partial \boldsymbol{\beta}}, \text{ and } \mathbf{G} = \frac{\partial \boldsymbol{\pi}(\boldsymbol{\beta})}{\partial \boldsymbol{\beta}}.$$

The $\mathbf{H}$ matrix can also be used to create residuals for marginals. A vector of simple residuals for marginals of any order can be defined as $\boldsymbol{e} = \mathbf{H}(\hat{\mathbf{p}} - \boldsymbol{\pi}(\hat{\boldsymbol{\beta}})) = \mathbf{H}\,\mathbf{u}$. As defined previously, $T = 2^q$ and g is the dimension of $\boldsymbol{\beta}$. If $\mathbf{H}$ contains $T - g - 1$ linearly independent rows corresponding to marginals from order 1 to q, then define the statistic

$$\chi^2_{[T-g-1]} = n\ \mathbf{u}'\mathbf{H}'\boldsymbol{\Omega}_e^{-1}\mathbf{H}\ \mathbf{u}. \tag{3.2}$$

Here, the statistic is evaluated at $\boldsymbol{\beta} = \hat{\boldsymbol{\beta}}$, where $\hat{\boldsymbol{\beta}}$ is now consistent and efficient for $\boldsymbol{\beta}$, such as the maximum likelihood estimator, and where $\boldsymbol{\Omega}_e = \mathbf{H}\boldsymbol{\Omega}_{\mathbf{u}}\mathbf{H}'$. With the added condition that the rows of $\mathbf{H}$ are linearly independent of the columns of the model matrix for $\boldsymbol{\pi}(\boldsymbol{\beta})$, $\chi^2_{[T-g-1]}$ can be shown to be equivalent to χ^2_{PF}. See Reiser (1996). To obtain orthogonal components, define the upper triangular matrix $\boldsymbol{F}$ such that $\boldsymbol{F}'\boldsymbol{\Omega}_e\boldsymbol{F} = \boldsymbol{I}$. $\boldsymbol{F} = (\boldsymbol{C}')^{-1}$, where $\boldsymbol{C}$ is the Cholesky factor of $\boldsymbol{\Omega}_e$. Then writing $\boldsymbol{\Omega}_e$ as $\boldsymbol{C}\boldsymbol{C}'$,

$$\begin{aligned}\chi^2_{PF} &= n\ \mathbf{u}'\mathbf{H}'(\hat{\boldsymbol{C}}')^{-1}\hat{\boldsymbol{C}}'(\hat{\boldsymbol{C}}\hat{\boldsymbol{C}}')^{-1}\hat{\boldsymbol{C}}(\hat{\boldsymbol{C}})^{-1}\mathbf{H}\ \mathbf{u}\\ &= n\ \mathbf{u}'\mathbf{H}'\widehat{\boldsymbol{F}}\widehat{\boldsymbol{F}}'\mathbf{H}\ \mathbf{u}\end{aligned}$$

where $\widehat{\boldsymbol{F}}$ and $\hat{\boldsymbol{C}}$ are the matrices $\boldsymbol{F}$ and $\boldsymbol{C}$ evaluated at $\boldsymbol{\beta} = \hat{\boldsymbol{\beta}}$. Premultiplication by $(\boldsymbol{C}')^{-1}$ orthonormalizes the matrix $\mathbf{H}$ relative to the matrix $D(\boldsymbol{\pi}) - \boldsymbol{\pi}\boldsymbol{\pi}' - \mathbf{G}(\mathbf{A}'\mathbf{A})^{-1}\mathbf{G}'$.

Let $\mathbf{H}^* = \boldsymbol{F}'\mathbf{H}$, then

$$\chi^2_{PF} = n\ \mathbf{u}'(\widehat{\mathbf{H}}^*)'\widehat{\mathbf{H}}^*\ \mathbf{u} \tag{3.3}$$

where $\widehat{\mathbf{H}}^* = \mathbf{H}^*(\hat{\boldsymbol{\beta}})$. Define $\hat{\boldsymbol{\gamma}} = n^{\frac{1}{2}} \widehat{\boldsymbol{F}}' \mathbf{H}\, \mathbf{u} = n^{\frac{1}{2}} \widehat{\mathbf{H}}^* \, \mathbf{u}$. Then,

$$\chi^2_{PF} = \hat{\boldsymbol{\gamma}}' \hat{\boldsymbol{\gamma}} = \sum_{k=1}^{k=T-g-1} \hat{\gamma}_k^2, \tag{3.4}$$

and the elements $\hat{\gamma}_k^2$ are orthogonal components of χ^2_{PF}. Since $\widehat{\mathbf{H}}^* \, \mathbf{u}$ statistics have asymptotic covariance matrix $\boldsymbol{F}' \boldsymbol{\Omega}_{\boldsymbol{e}} \boldsymbol{F} = \boldsymbol{I}_{T-g-1}$, the elements $\hat{\gamma}_k^2$ are asymptotically independent χ_1^2 random variables, assuming a consistent estimator for $\boldsymbol{\pi}(\boldsymbol{\beta})$ and $\boldsymbol{\Sigma}_{\mathbf{e}}$ where $\widehat{\boldsymbol{\Sigma}}_{\mathbf{e}} = n^{-1} \hat{\boldsymbol{\Omega}}_{\boldsymbol{e}}$. The asymptotic approximation may not be valid for components from a sparse higher order marginal table (Reiser and VandenBerg 1994).

Then for binary cross-classified variables, define $GFfit_{\perp}^{(ij)} = \hat{\gamma}_k^2$, where $k = q+1, q+2, ..., q(q+1)/2; i = 1, 2, ..., q-1; j = i+1, ..., q$. When variables have $c \geq 2$ categories, $GFfit_{\perp}^{(ij)}$ is a sum of $(c-1)^2$ orthogonal components (Reiser 2008). $GFfit_{\perp}^{(ij)}$ is an extended version of the $GFfit^{(ij)}$ statistic from Cagnone and Mignani (2007). $GFfit_{\perp}^{(ij)}$ statistics have a sequential property due to the application of the Cholesky factor.

Since $GFfit_{\perp}^{(ij)}$ statistics are asymptotically independent χ_1^2 random variates, they can be summed to form the statistic $X^2_{[2]}$ which is distributed asymptotically as Chi-square (Reiser 2008). This statistic can be used for a more omnibus test *focused* on the second-order marginals with null hypothesis $H_0 \colon \mathbf{H}_{[2]}\boldsymbol{\pi} = \mathbf{H}_{[2]}\boldsymbol{\pi}(\boldsymbol{\beta})$. If this null hypothesis is rejected, then it follows that the null hypothesis $H_0 \colon \boldsymbol{\pi} = \boldsymbol{\pi}(\boldsymbol{\beta})$ should be rejected. Since $X^2_{[2]}$ is calculated from second-order marginals, the asymptotic Chi-square distribution can be expected to be valid even when the full cross-classified table is sparse (Reiser 1996).

3.2 *Adjusted Residuals*

The adjusted residual for second-order marginal i, j is

$$Z_{ij} = \frac{n^{\frac{1}{2}} e^{(k)}}{\hat{\sigma}_{e^{(k)}}}, \tag{3.5}$$

where $k = 1, 2, ..., \binom{q}{2}$ and corresponds to item pair i, j, $e^{(k)}$ is an element of $\boldsymbol{e}$ and $\hat{\sigma}_{e^{(k)}}$ is the square root of the diagonal elements of $\widehat{\boldsymbol{\Sigma}}_{\mathbf{e}}$. Adjusted residuals on marginal tables for binary variables were developed by Reiser (1996). $\hat{\gamma}_k$, defined above, is essentially an orthogonalized version of this adjusted residual.

3.3 The $\bar{\chi}^2_{ij}$ Statistic

The authors of Liu and Maydeu-Olivares (2014) proposed a mean and variance adjusted Chi-square statistic, $\bar{\chi}^2_{ij}$, for the bivariate distribution of variables i, j within a large table. Consider the case where Pearson's Chi-square is applied to a bivariate subtable,

$$\chi^2_{ij} = n(\mathbf{p}_{ij} - \hat{\boldsymbol{\pi}}_{ij})'\mathbf{D}^{-1}_{ij}(\mathbf{p}_{ij} - \hat{\boldsymbol{\pi}}_{ij}) \tag{3.6}$$

where $\mathbf{D}_{ij} = diag(\pi_{ij})$ is a diagonal matrix of the bivariate probabilities. If $\boldsymbol{\beta}$ is estimated on the full table, χ^2_{ij} is not distributed asymptotically as Chi-square and has an unknown distribution function. But a mean and variance adjusted statistic has asymptotic distribution that can be well approximated by the Chi-square distribution. Then, the mean and variance adjusted statistic, $\bar{\chi}^2_{ij}$, is defined as

$$\bar{\chi}^2_{ij} = 2\frac{\hat{\mu}_1}{\hat{\mu}_2}\chi^2_{ij}, \tag{3.7}$$

where the definitions of the two asymptotic moments (μ_1, μ_2) are given in the online supplement. $\bar{\chi}^2_{ij}$ has an approximate reference Chi-square distribution with degrees of freedom $a = \frac{2\hat{\mu}_1^2}{\hat{\mu}_2}$, but the joint distribution of a set of $\bar{\chi}^2_{ij}$ for a model on a cross-classified table is unknown.

4 Simulation Studies

4.1 Type I Error Study

The first simulation examined Type I error for the lack-of-fit diagnostics reviewed above using a continuous latent factor model for categorical variables known as the two-parameter logistic (2PL) item response model (Bock and Lieberman 1975). In this model, $\boldsymbol{\beta} = (\boldsymbol{\beta}_0, \boldsymbol{\beta}_1)'$, where $\boldsymbol{\beta}_0$ is a vector of intercepts, and $\boldsymbol{\beta}_1$ is a vector of slopes. In this study with $q = 8$ manifest variables, $\boldsymbol{\beta}_1' = (0.1, 0.1, 0.1, 0.9, 0.9, 0.9, 0.2, 0.2)$. Results shown below are for intercepts symmetric around zero, $\boldsymbol{\beta}_0' = (-2.0, -1.5, -1.0, -0.5, 0.5, 1.0, 1.5, 2.0)$. Results for simulations with asymmetric intercepts are shown in the online supplement. Using Monte Carlo methods, 1,000 data sets were generated for each setting. Empirical Type I error rates of the individual lack-of-fit diagnostics were calculated. Since an individual $GFfit_{\perp}^{(ij)}$ is distributed approximately as Chi-square with one degree of freedom, to calculate the empirical Type I error rate for each $GFfit_{\perp}^{(ij)}$, the sum of the number of cases that exceed the Chi-square critical value (at 5% significance level) with one degree of freedom was divided by the number of data sets. A similar process was used to calculate the

Type I error rates of the adjusted residual and $\bar{\chi}^2_{ij}$. This simulation was repeated for sample sizes 300 and 500, as is common in the literature on lack-of-fit diagnostics (Cagnone and Mignani 2007; Liu and Maydeu-Olivares 2014; Reiser 1996; Reiser et al. 2022). The $GFfit_{\perp}^{(ij)}$ statistics were calculated using an orthogonal regression given in Reiser (2008).

Table 1 below indicates the empirical Type I error rates for $q = 8$ manifest variables for the symmetric intercept setting. The Type I error rates outside of the Monte Carlo error interval $0.05 \pm \sqrt{0.05(0.95)/1000} = (0.0365, 0.0635)$ are bolded. When $n = 300$, Type I error rates related to $GFfit_{\perp}^{(ij)}$ (4,5) and (5,6) were outside the Monte Carlo error interval. Given that there are twenty-eight individual $GFfit_{\perp}^{(ij)}$, it is possible that one or two components may randomly fall slightly outside the Monte Carlo error interval. However, error rates for five $\bar{\chi}^2_{ij}$ and four adjusted residuals were outside the Monte Carlo error interval. This suggests, when $n = 300$, $GFfit_{\perp}^{(ij)}$ has a better Type I error rate compared to $\bar{\chi}^2_{ij}$ and adjusted residuals for eight manifest variables with a symmetric intercept model.

When $n = 500$, all of the error rates for $GFfit_{\perp}^{(ij)}$ were inside the Monte Carlo interval, while $\bar{\chi}^2_{ij}$ and adjusted residuals each had two error rates outside the Monte Carlo interval. Similar results for Type I error were found in simulations using 15 variables, as shown in the online supplement.

4.2 Estimated Mean and Variance of the Statistics

Since an individual $GFfit_{\perp}^{(ij)}$ is distributed approximately as χ^2_1, it is expected that the mean be close to 1 and SD be close to $\sqrt{2}$. Results in Table 2 suggests that $GFfit_{\perp}^{(ij)}$ has mean and SD close 1 and $\sqrt{2}$, respectively. For the adjusted residuals, the mean should be close to 0 and the SD should be close to 1. $\bar{\chi}^2_{ij}$ should have an approximate Chi-square distribution, but the degrees of freedoms may not be the same for each pair. If a statistic has a larger empirical variance, in general larger than 2 times the mean for statistics distributed as Chi-squared, then it is an extremely undesirable feature in that statistic in terms of real-world applications, because even in correctly specified models, one may observe extremely large values of such a statistic in applications, which may lead researchers to believe that the model grossly misfits one or more pairs. Table 2 below indicates when $n = 300$, $GFfit_{\perp}^{(ij)}$ has similar results to $\bar{\chi}^2_{ij}$ for mean and SD and better results compared to adjusted residuals. High variance for the version of $GFfit^{(ij)}$ in Cagnone and Mignani (2007) was reported by Liu and Maydeu-Olivares (2014). This high variance will result from the direct application of a matrix inverse to $\widehat{\mathbf{\Sigma}}_{\mathbf{e}}$, an approach which lacks sufficient numerical stability. Calculation of $GFfit^{(ij)}$ using an orthogonal regression, as in the present study, has high numerical stability.

Table 1 Type I error study for 2PL latent variable model, symmetric intercepts

	$n = 300$			$n = 500$		
Pair (i,j)	$GFfit_{\perp}^{(ij)}$	Adj. Residuals	$\bar{\chi}_{ij}^2$	$GFfit_{\perp}^{(ij)}$	Adj. Residuals	$\bar{\chi}_{ij}^2$
(1, 2)	0.046	0.055	0.052	0.052	0.059	0.056
(1, 3)	0.048	0.048	0.047	0.044	0.046	0.046
(1, 4)	0.044	0.057	0.054	0.051	0.051	0.047
(1, 5)	0.044	**0.034**	**0.03**	0.042	0.043	0.043
(1, 6)	0.049	0.048	0.044	0.053	0.049	0.047
(1, 7)	0.051	0.063	0.060	0.043	0.045	0.045
(1, 8)	0.057	0.053	0.052	0.051	0.053	0.053
(2, 3)	0.051	0.055	0.054	0.041	0.042	0.043
(2, 4)	0.039	0.046	0.049	0.038	0.047	0.046
(2, 5)	0.043	0.054	0.052	0.049	0.050	0.050
(2, 6)	0.052	0.063	0.059	0.048	0.042	0.042
(2, 7)	0.043	0.059	0.060	0.048	0.049	0.047
(2, 8)	0.047	0.048	0.048	0.057	0.053	0.054
(3, 4)	0.050	0.058	0.058	0.05	0.052	0.051
(3, 5)	0.042	0.038	0.038	0.043	0.044	0.042
(3, 6)	0.049	0.060	0.056	0.051	0.046	0.046
(3, 7)	0.043	0.048	0.049	0.056	0.050	0.048
(3, 8)	0.041	0.043	0.043	0.047	0.039	0.040
(4, 5)	**0.074**	**0.080**	**0.079**	0.064	**0.070**	**0.069**
(4, 6)	0.062	**0.079**	**0.077**	0.057	**0.068**	**0.067**
(4, 7)	0.037	0.054	0.052	0.037	0.051	0.049
(4, 8)	0.050	0.042	0.042	0.048	0.042	0.042
(5, 6)	**0.070**	**0.074**	**0.073**	0.062	0.063	0.064
(5, 7)	0.039	0.044	0.043	0.039	0.037	0.038
(5, 8)	0.052	0.050	0.052	0.037	0.037	0.037
(6, 7)	0.045	0.045	0.048	0.054	0.048	0.050
(6, 8)	0.037	0.044	0.044	0.049	0.037	0.038
(7, 8)	0.052	0.040	**0.036**	0.041	0.037	0.040

1000 samples; 975 ($n = 300$), 998 ($n = 500$) convergence

4.3 Power Study for Eight Variables

A power study was conducted by using Monte Carlo methods to generate 1,000 data sets from a model with two continuous latent factors. Intercepts and slopes for the first factor were the same as in the Type I error study. For the second factor, $\boldsymbol{\beta}_2'$ = (0,0,0,1,1,1,1,1). Higher power is expected for lack-of-fit diagnostics related to variables 1, 2 and 3. Data sets were fit using a false model with one continuous latent

Table 2 Mean and SD for 2PL latent variable model, symmetric intercepts, eight variables

	$n = 300$						$n = 500$					
	$GFfit_{\perp}^{(ij)}$		Adj. Res.		$\bar{\chi}_{ij}^2$		$GFfit_{\perp}^{(ij)}$		Adj. Res.		$\bar{\chi}_{ij}^2$	
Pair(i,j)	Mean	SD	Mean	SD	Mean	SD	Mean	SD	Mean	SD	Mean	SD
(1, 2)	0.99	1.42	0.14	1.24	0.98	1.42	1.09	1.55	–0.02	1.13	1.14	1.95
(1, 3)	1.06	1.62	0.05	1.24	0.96	1.29	0.96	1.36	0.04	1.09	1.01	1.82
(1, 4)	1.16	2.08	0.44	3.13	0.94	1.37	0.97	1.33	–0.01	0.99	1.02	1.98
(1, 5)	1.03	1.64	–0.83	3.49	0.96	1.34	0.98	1.40	0.00	0.98	1.00	1.73
(1, 6)	1.12	1.57	–0.71	2.78	0.93	1.25	1.02	1.50	0.02	1.01	1.03	1.57
(1, 7)	1.02	1.47	–0.10	1.27	0.97	1.25	1.03	1.57	0.00	1.03	1.07	1.57
(1, 8)	1.01	1.41	–0.15	1.32	1.02	1.67	1.00	1.44	0.00	1.02	1.07	1.69
(2, 3)	0.97	1.33	0.01	1.03	0.96	1.42	0.99	1.35	–0.03	0.99	0.98	1.34
(2, 4)	1.00	1.48	0.45	2.83	0.87	1.24	0.94	1.33	0.01	0.97	0.94	1.34
(2, 5)	1.04	1.67	–0.78	3.56	1.00	1.57	0.93	1.30	–0.02	1.00	0.99	1.38
(2, 6)	1.06	1.56	–0.60	2.57	0.92	1.27	1.02	1.43	0.03	0.97	0.94	1.29
(2, 7)	1.06	1.47	–0.03	1.12	0.99	1.29	1.00	1.43	0.01	1.01	1.02	1.37
(2, 8)	1.02	1.45	–0.13	1.16	1.01	1.41	0.99	1.25	0.02	1.00	0.99	1.44
(3, 4)	1.04	1.91	0.31	2.46	1.05	1.39	0.99	1.39	–0.04	1.00	1.00	1.41
(3, 5)	1.05	1.39	–0.68	2.95	0.98	1.37	0.94	1.34	0.03	0.97	0.94	1.38
(3, 6)	1.09	1.51	–0.57	2.44	0.97	1.34	1.00	1.45	0.01	0.99	0.98	1.43
(3, 7)	1.08	1.48	–0.07	1.15	0.99	1.33	1.03	1.47	0.00	1.00	1.00	1.45
(3, 8)	1.01	1.48	–0.11	1.10	0.96	1.29	0.98	1.28	0.03	1.01	1.02	1.96
(4, 5)	1.21	2.36	–2.30	**8.57**	0.99	1.52	1.06	2.01	0.03	1.09	1.17	2.19
(4, 6)	1.15	1.80	–1.39	5.34	0.92	1.38	0.97	1.35	0.08	0.99	0.99	1.47
(4, 7)	0.98	1.51	–0.06	2.03	0.98	1.30	1.01	1.39	0.00	1.00	0.99	1.35
(4, 8)	1.10	1.51	–0.07	1.89	0.91	1.28	1.00	1.40	0.05	0.97	0.94	1.27
(5, 6)	1.14	1.73	–2.79	**9.15**	1.05	1.44	1.07	1.54	0.00	1.05	1.11	2.03
(5, 7)	0.98	1.35	–0.48	2.35	0.92	1.34	0.99	1.45	–0.02	0.98	0.95	1.34
(5, 8)	1.17	1.88	–0.41	2.18	0.92	1.30	0.99	1.43	–0.03	0.97	0.94	1.23
(6, 7)	1.18	2.02	–0.47	2.45	1.05	1.41	0.97	1.43	0.05	1.02	1.03	1.50
(6, 8)	1.02	1.49	–0.41	2.22	0.99	1.38	0.97	1.31	–0.04	1.00	1.00	1.41
(7, 8)	1.09	1.55	–0.08	1.16	1.04	1.38	0.98	1.36	0.01	0.97	0.94	1.36

factor. The process was repeated for $n = 300$ and $n = 500$. Asymptotic and empirical power comparisons are given in Table 3. Asymptotic power is displayed only for $GFfit_{\perp}^{(ij)}$. By examining the highlighted values, it is clear that the empirical power of second-order $GFfit_{\perp}^{(ij)}$ $(1, 2)$, $(1, 3)$ and $(2, 3)$ are substantially higher compared to other components. Thus, these second-order $GFfit_{\perp}^{(ij)}$ were successful in detecting the source of a poorly fit model. Empirical power for the three diagnostics was close for variable pairs (1, 2) and (1, 3), but $GFfit_{\perp}^{(ij)}$ had higher power for pair (2, 3). These power results should be considered in the context that $GFfit_{\perp}^{(ij)}$ has better Type I error performance when $n = 300$. Empirical power results for $GFfit_{\perp}^{(ij)}$ were

Table 3 Asymptotic and empirical power comparison for 2PL latent variable model, eight variables

	$n = 300$				$n = 500$			
Pair (i,j)	$GFfit_{\perp}^{(ij)}$	Adj. Resi.	$\bar{\chi}_{ij}^{2}$	Asy. power*	$GFfit_{\perp}^{(ij)}$	Adj. Resi.	$\bar{\chi}_{ij}^{2}$	Asy. power[a]
(1, 2)	**0.584**	**0.601**	**0.599**	**0.660**	**0.835**	**0.826**	**0.824**	**0.865**
(1, 3)	**0.543**	**0.553**	**0.553**	**0.609**	**0.814**	**0.781**	**0.781**	**0.823**
(1, 4)	0.060	0.064	0.063	0.051	0.054	0.080	0.081	0.051
(1, 5)	0.049	0.058	0.055	0.052	0.056	0.080	0.081	0.053
(1, 6)	0.048	0.064	0.066	0.054	0.045	0.074	0.075	0.057
(1, 7)	0.054	0.047	0.049	0.050	0.051	0.068	0.069	0.050
(1, 8)	0.045	0.065	0.065	0.050	0.052	0.060	0.060	0.050
(2, 3)	**0.610**	**0.562**	**0.562**	**0.654**	**0.852**	**0.803**	**0.803**	**0.860**
(2, 4)	0.048	0.072	0.072	0.051	0.059	0.064	0.066	0.052
(2, 5)	0.043	0.062	0.067	0.052	0.060	0.073	0.072	0.053
(2, 6)	0.067	0.056	0.055	0.054	0.060	0.075	0.074	0.057
(2, 7)	0.044	0.059	0.057	0.050	0.042	0.054	0.055	0.050
(2, 8)	0.056	0.054	0.056	0.050	0.058	0.061	0.060	0.050
(3, 4)	0.056	0.064	0.064	0.051	0.067	0.062	0.065	0.052
(3, 5)	0.056	0.053	0.054	0.052	0.061	0.065	0.065	0.053
(3, 6)	0.071	0.048	0.049	0.054	0.070	0.075	0.075	0.057
(3, 7)	0.061	0.056	0.055	0.050	0.052	0.064	0.064	0.050
(3, 8)	0.053	0.066	0.064	0.050	0.040	0.065	0.066	0.050
(4, 5)	0.050	0.07	0.073	0.050	0.052	0.063	0.064	0.050
(4, 6)	0.051	0.073	0.071	0.050	0.053	0.076	0.076	0.050
(4, 7)	0.043	0.058	0.06	0.050	0.055	0.055	0.055	0.050
(4, 8)	0.051	0.063	0.062	0.050	0.040	0.059	0.057	0.050
(5, 6)	0.059	0.079	0.077	0.056	0.046	0.063	0.062	0.050
(5, 7)	0.045	0.061	0.061	0.050	0.053	0.052	0.050	0.050
(5, 8)	0.053	0.052	0.048	0.050	0.051	0.049	0.049	0.050
(6, 7)	0.049	0.056	0.057	0.050	0.054	0.062	0.062	0.050
(6, 8)	0.050	0.070	0.064	0.050	0.050	0.061	0.060	0.050
(7, 8)	0.051	0.070	0.072	0.050	0.049	0.057	0.058	0.050

[a]Asymptotic power was calculated only for the orthogonal components
1000 samples; 984 ($n = 300$), 999 ($n = 500$) convergence

somewhat lower compared to asymptotic power results when $n = 300$, indicating that when sample size is smaller the empirical distribution may not be close to the theoretical power function.

From the results in Table 3, it is clear that the empirical power will increase with the sample size. When $n = 500$, $GFfit^{(ij)}$ had higher power for all three variable pairs and substantially higher power for pair (2, 3). An important advantage in applications is that $GFfit_{\perp}^{(ij)}$ statistics have an independence property. Also when

$n = 500$, empirical power results and asymptotic power results for $GFfit_{\perp}^{(ij)}$ were fairly close, indicating that the empirical distribution approaches the theoretical distribution. Similar results for power were found in simulations using 15 variables.

5 Application

The Epidemiologic Catchment Area (ECA) (United States Department of Health and Human Services 1985) program of research was initiated in response to the 1977 report of the U.S. President's Commission on Mental Health. The purpose was to collect data on the prevalence and incidence of mental disorders and on the use of and need for services by the mentally ill. For this application, eight items related to the psychiatric condition known as simple phobias were chosen from the ECA study to analyze as a real-world application. The data set was limited to Johns Hopkins (Baltimore, MD, U.S.A.) area. The selected items are (1) fear of heights, (2) fear of closed places, (3) fear of speaking in front of close friends, (4) fear of speaking to strangers, (5) storms, (6) water, (7) spiders and (8) fear of harmless animals. There were 3,316 observations related to these specifications. Each variable has two categories: 'yes' or 'no'. Thus, there are $2^8 = 256$ response patterns. However, as most of the answers are 'no', many response patterns have a cell count less than five, and 165 response patterns are empty.

A model with one latent factor was fitted to the data. For this model, $\chi^2_{PF} = 488.95$ on 239 degrees of freedom. Since the overall table is sparse, the Chi-square approximation for χ^2_{PF} may not be valid. However, the bivariate tables are not sparse, and $\chi^2_{[2]} = 108.53$ on 28 degrees of freedom with *p-value* $= 1.51E - 10$, indicating that the model with one factor is not a good fit. $GFfit_{\perp}^{(ij)}$ was used to identify lack-of-fit. When a large number of components is produced, a multiple decision rule should be used to determine which components are significantly large. Because the $GFfit_{\perp}^{(ij)}$ are independent random variates, it is possible to take advantage of the False Discovery Rate (FDR) (Benjamini et al. 1995) procedure for independent tests to maintain Type I error rate. FDR adjustment is not valid for $\bar{\chi}^2_{ij}$ because it has unknown joint distribution, so a more conservative method such as the Benjamini and Yekutieli (2001) procedure would be needed to maintain Type I error for $\bar{\chi}^2_{ij}$.

The lack-of-fit diagnostics for second-order marginals are shown in Table 4 along with the raw *p-values* and the adaptive FDR *p-values* for $GFfit_{\perp}^{(ij)}$. In the table, $GFfit_{\perp}^{(ij)}$ (1, 8), (3, 4), (3, 7), (3, 8) and (6, 7) have significant FDR *p-values* indicating that these pairs of variables have associations not explained by the model with one factor. Results related to $GFfit_{\perp}^{(ij)}$, adjusted residual and $\bar{\chi}^2_{ij}$ are consistent with each other. Further, variable 3, 'fear of speaking in front of close friends' appears in three of these large components. 'Fear of speaking' may represent more than a simple phobia.

Table 4 $GFfit_{\perp}^{(ij)}$, adjusted residuals and $\bar{\chi}_{ij}^2$ for ECA study

Pair (i,j)	$GFfit_{\perp}^{(ij)}$	Adj. Residual	$\bar{\chi}_{ij}^2$	$GFfit_{\perp}^{(ij)}$ Raw $P-value$	$GFfit_{\perp}^{(ij)}$ FDR $P-value$
(1, 2)	1.637	1.153	1.285	0.200	0.423
(1, 3)	1.504	–1.350	1.905	0.220	0.423
(1, 4)	3.702	–2.289	5.019	0.054	0.176
(1, 5)	0.277	0.658	0.293	0.598	0.698
(1, 6)	0.768	0.617	0.256	0.380	0.560
(1, 7)	1.204	–1.220	3.802	0.272	0.438
(1, 8)	11.825	–3.328	11.132	0.001	**0.008**
(2, 3)	0.056	–0.566	0.278	0.812	0.842
(2, 4)	0.508	–1.967	3.506	0.475	0.614
(2, 5)	0.172	–0.905	0.903	0.677	0.731
(2, 6)	3.633	–0.109	0.037	0.056	0.176
(2, 7)	0.568	–2.752	9.193	0.450	0.614
(2, 8)	0.445	–1.378	1.731	0.504	0.614
(3, 4)	32.109	4.752	24.251	1.45E-08	**4.08E-07**
(3, 5)	1.460	–2.387	5.730	0.226	0.423
(3, 6)	0.0005	–1.502	2.262	0.982	0.982
(3, 7)	10.984	–3.543	13.571	0.001	**0.008**
(3, 8)	10.077	–1.844	3.213	0.001	**0.010**
(4, 5)	1.872	–3.238	10.168	0.171	0.399
(4, 6)	1.281	–1.523	2.104	0.257	0.438
(4, 7)	2.706	–2.536	6.602	0.099	0.279
(4, 8)	2.555	–1.789	2.823	0.109	0.279
(5, 6)	3.910	–1.503	2.734	0.047	0.176
(5, 7)	0.170	0.685	0.501	0.679	0.731
(5, 8)	4.437	0.221	0.115	0.035	0.164
(6, 7)	9.014	–2.855	11.429	0.002	**0.014**
(6, 8)	0.475	–2.101	4.335	0.490	0.614
(7, 8)	1.158	2.335	6.135	0.281	0.438

6 Conclusions

$GFfit_{\perp}^{(ij)}$ statistics are asymptotic independent Chi-square variates obtained as orthogonal components of Pearson's χ_{PF}^2, and $GFfit_{\perp}^{(ij)}$ is a powerful diagnostic to detect the source of lack of fit when a hypothesized model does not fit a table of cross-classified binary variables. Power calculations for a latent variable model show that $GFfit_{\perp}^{(ij)}$ has higher power than adjusted residuals and $\bar{\chi}_{ij}^2$. Simulation results show that $GFfit_{\perp}^{(ij)}$ has good Type I error performance even if the joint frequencies in full 2^q table are sparse and that $GFfit_{\perp}^{(ij)}$ is computationally stable when calculated with an orthogonal regression.

References

Benjamini, Y., & Hochberg, Y. (1995). Controlling the false discovery rate: A practical and powerful approach to multiple testing. *Journal of the Royal Statistical Society: Series B (Methodological), 57*(1), 289–300.

Benjamini, Y., & Yekutieli, D. (2001). The control of the false discovery rate in multiple testing under dependency. *Annals of Statistics, 29*(4), 1165–1188.

Bock, R. D., & Lieberman, M. (1975). Fitting a response model for *n* dichotomously score items. *Psychometrika, 40*, 5–32.

Cagnone, S., & Mignani, S. (2007). Assessing the goodness of fit for a latent variable model for ordinal data. *Metron*, LXV, 337–361.

Fisher, R. A. (1924). The conditions under which chi square measures the discrepancy between observation and hypothesis. *Journal of the Royal Statistical Society, 87*, 19–43.

Lancaster, H. (1969). *The chi-squared distribution* (Vol. 40, pp. 6667) Wiley.

Liu, Y., & Maydeu-Olivares, A. (2014). Identifying the source of misfit in item response theory models. *Multivariate Behavioral Research, 49*(4), 354–371.

Rayner, J. C. W., & Best, D. J. (1989). *Smooth tests of goodness of fit*. New York: Oxford.

Reiser, M. (1996). Analysis of residuals for the multionmial item response model. *Psychometrika, 61*(3), 509–528.

Reiser, M. (2008). Goodness-of-fit testing using components based on marginal frequencies of multinomial data. *British Journal of Mathematical and Statistical Psychology, 61*(2), 331–360.

Reiser, M., & VandenBerg, M. (1994). Validity of the Chi-square test in dichotomous variable factor analysis when expected frequencies are small. *British Journal of Mathematical and Statistical Psychology, 47*, 85–107.

Reiser, M., Cagone, S., & Zhu, J. (2022). An extended GFfit statistic defined on orthogonal components of Pearson's chi-square. *Psychometrika*. https://doi.org/10.1007/s11336-022-09866-8

United States Department of Health and Human Services, National Institute of Mental Health. (1985). *Epidemiological Catchment Area (ECA) Survey of Mental Disorders, Wave I (Household), 1980–1985: [United States]*. Rockville, MD: [producer], 1985. Ann Arbor, MI: Inter-university Consortium for Political and Social Research [distributor], 1991. https://doi.org/10.3886/ICPSR08993.v1.

Robust Response Transformations for Generalized Additive Models via Additivity and Variance Stabilization

Marco Riani, Anthony C. Atkinson, and Aldo Corbellini

Abstract The AVAS (Additivity And Variance Stabilization) algorithm of Tibshirani provides a non-parametric transformation of the response in a linear model to approximately constant variance. It is thus a generalization of the much used Box-Cox transformation. However, AVAS is not robust. Outliers can have a major effect on the estimated transformations both of the response and of the transformed explanatory variables in the Generalized Additive Model (GAM). We describe and illustrate robust methods for the non-parametric transformation of the response and for estimation of the terms in the model and report the results of a simulation study comparing our robust procedure with AVAS. We illustrate the efficacy of our procedure through a simulation study and the analysis of real data.

Keywords Augmented star plot · AVAS · Backfitting · Forward search · Heatmap · Outlier detection · Robust regression

1 Introduction

The nonlinear parametric transformation of response variables are a common practice in regression problems, for example, logarithms of survival times. Tibshirani (1988) used smoothing techniques to provide non-parametric transformations of the response together with transformations of the explanatory variables, a procedure he

M. Riani (✉) · A. Corbellini
Dipartimento di Scienze Economiche e Aziendale and Interdepartmental Centre for Robust Statistics, Università di Parma, 43100 Parma, Italy
e-mail: mriani@unipr.it
URL: http://www.riani.it

A. Corbellini
e-mail: aldo.corbellini@unipr.it

A. C. Atkinson
The London School of Economics, London WC2A 2AE, UK
e-mail: a.c.atkinson@lse.ac.uk

L. Grilli et al. (eds.), *Statistical Models and Methods for Data Science*, Studies in Classification, Data Analysis, and Knowledge Organization,
https://doi.org/10.1007/978-3-031-30164-3_12

called AVAS (additivity and variance stabilization). The resulting model is a generalized additive model (GAM) with a response transformed to approximate constant variance. Tibshirani's work can be seen as a non-parametric extension of the power transformation family of Box and Cox (1964) in which the goals are the stabilization of error variance and the approximate normalization of the error distribution, hopefully combined with an additive model. It also extends the parametric transformation of explanatory variables of Box and Tidwell (1962). A discussion of the relationship of AVAS to the Box-Cox transformation is in Hastie and Tibshirani (1990, Chap. 7).

Tibshirani's AVAS is not robust with respect to outliers. Our main purpose is to provide a robust version of his work, which, for obvious reasons, we call RAVAS. In developing our procedure we made four important improvements to the original AVAS. Like robustness, these are available as options. Thus, RAVAS can be used for fitting a response transformed GAM when robustness is not an issue, or for fitting a GAM without response transformation.

Section 2 introduces the generalized additive model and the associated backfitting algorithm for estimation of the transformations of the explanatory variables, which uses a smoothing algorithm. The AVAS procedure and the associated numerical variance stabilization transformation are described in Sects. 2.3 and 2.4. Section 3 outlines the various forms of robust regression that are available in our algorithm and describes the resulting outlier detection procedures. The purpose is to provide an outlier free subset of the data for transformation and smoothing. An outline of our improvements to AVAS is in Sect. 4. Appreciably more detail of these is provided in Riani et al. (2023) as well as further data analyses. Section 5 presents the results of a simulation study comparing some properties of AVAS and RAVAS in the presence of outliers: the mean squared error of parameter estimates, the power of detection of outliers, (just for RAVAS) and the number of numerical iterations of the two algorithms required for convergence. The performance of AVAS is severely degraded by the presence of outliers. The last two sections present a data analysis, which makes use of the augmented star plot as a guide to the choice of options in the estimation process and includes a comparison of the choices using a heatmap of correlations. The purpose of the paper is both to introduce the MATLAB program we have written for this form of robust data analysis and to illustrate some of its properties.

2 Generalized Additive Models and the Structure of AVAS

2.1 Introduction

The generalized additive model (GAM) has the form

$$g(Y_i) = \beta_0 + \sum_{j=1}^{p} f_j(X_{ij}) + \epsilon. \tag{1}$$

The functions f_j are unknown and are, in general, found by the use of smoothing techniques. A monotonicity constraint can be applied. If the response transformation or link function g is unknown, it is restricted to be monotonic, but scaled to satisfy the technically necessary constraint that $\text{var}\{g(Y)\} = 1$. In the fitting algorithm, the transformed responses are scaled to have mean zero; the constant β_0 can therefore be ignored. The observational errors are assumed to be independent and additive with constant variance. The performance of fitted models is compared by use of the coefficient of determination R^2. Since the f_j are estimated from the data, the traditional assumption of linearity in the explanatory variables is avoided. However, the GAM retains the assumption that explanatory variable effects are additive. Buja et al. (1989) describe the background and early development of this model.

2.2 Backfitting

For the moment we assume that the response transformation $g(Y)$ is known. The backfitting algorithm, described in Hastie and Tibshirani (1990, p. 91), is used to fit a GAM. The algorithm proceeds iteratively using residuals when one explanatory variable in turn is dropped from the model.

With $g(y)$ the $n \times 1$ vector of transformed responses, let $e_{(j)}$ be the vector of residuals when $f_j(x_j)$ is removed from the model without any refitting. Then

$$e_{(j)} = g(y) - \sum_{k \neq j = 1}^{p} f_k(x_k). \tag{2}$$

The new value of $f_j(.)$ depends on ordered values of $e_{(j)}$ and x_j. Let the ordered values of x_j be $x_{s,j}$. The residuals $e_{(j)}$ are sorted in the same way to give the new order $e_{s,(j)}$. Within each iteration each explanatory variable is dropped in turn; $j = 1, \ldots, p$. The iterations continue until the change in the value of R^2 is less than a specified tolerance.

For iteration l the vector of sorted residuals for x_j is $e^l_{(j)}$. The new estimate of $f_j^{(l+1)}$ is

$$f_{s,j}^{(l+1)} = S\left\{e^l_{s,(j)}, x_{s,j}\right\}. \tag{3}$$

The function S depends on the constraint imposed on the transformation of variable j. If the transformation can be non-monotonic, S denotes a smoothing procedure. As does Tibshirani (1988), we use the supersmoother (Friedman and Stuetzle 1982), a non-parametric estimator based on local linear regression with adaptive bandwidths. Monotonic transformations using isotonic regression are also an optional possibility (Barlow et al. 1972).

The backfitting algorithm is not invariant to the permutation of order of the variables inside matrix X, with high collinearity between the explanatory variables causing slow convergence of the algorithm: the residual sum of squares can change very little between iterations. Our option orderR2, Sect. 4.1.4, attempts a solution to this problem by reordering the variables in order of importance.

2.3 The AVAS Algorithm

In this section, we present the structure of the AVAS algorithm of Tibshirani (1988). The variance stabilizing transformation used to estimate the response transformation is outlined in Sect. 2.4

Our RAVAS algorithm has a similar structure to that given here, made more elaborate by the requirements of robustness and the presence of options. In this description of the algorithm ty and tX are transformed values of y and X.

1. *Initialize Data*. Standardize response y so that $\overline{ty} = 0$ and $\text{var}(ty) = 1$, where var is the maximum likelihood biased estimator of variance. Centre each column of the X matrix so that $\overline{tX}_j = 0, j = 1, \ldots, p)$.
2. *Initial call to 'Inner Loop'* to find initial GAM using ty and tX; calculates initial value of the coefficient of determination, R^2. Set convergence conditions on number of iterations and value of R^2.
3. *Main (Outer) Loop*. Given values of ty and tX at each iteration the outer loop finds numerically updated values of the transformed response. Given the newly transformed response, updated transformed explanatory variables are found through the call to the backfitting algorithm (*inner loop*). In our version iterations continue until a stopping condition on R^2 is verified or until a maximum number of iterations has been reached.

2.4 The Numerical Variance Stabilizing Transformation

We first consider the case of a random variable Y with known distribution for which $\text{E}(Y) = \mu$ and $\text{var}(Y) = V(\mu)$. We seek a transformation $ty = h(y)$ for which the variance is, at least approximately, independent of the mean. Then Taylor series expansion of $h(y)$ leads to $\text{var}(Y) \approx V(\mu)\{h'(\mu)\}^2$. For a general distribution $h(y)$ is then a solution of the differential equation $\text{d}g/\text{d}\mu = C/\sqrt{V(\mu)}$. For random variables standardized, as are the values ty, to have unit variance, $C = 1$ the variance stabilizing transformation is

$$h(t) = \int^t 1/\sqrt{V(u)}du. \tag{4}$$

In the AVAS algorithm for data, $1/\sqrt{V(u)}$ is estimated by the vector of the reciprocals of the absolute values of the smoothed residuals sorted using the ordering based on fitted values of the model. There are n integrals, one for each observation. The range of integration for observation i goes from the smallest fitted value to the old transformed value $\widehat{ty}_i, i = 1, \ldots, n$. The computation of the n integrals uses the trapezoidal rule and is outlined in Sect. 4.2. Since the transformation is the sum of an increasing number of non-negative elements, monotonicity is assured. The logged residuals in the estimation of the variance function are smoothed using the running line smoother of Hastie and Tibshirani (1986).

3 Robustness and Outlier Detection

3.1 Robust Regression

We robustify our transformation method through the use of robust regression to replace least squares. The examples in this paper have been calculated using *Adaptive Hard Trimming*. In the Forward Search (FS), the observations are hard trimmed, the amount of trimming being determined by the choice of the trimming parameter h, the value of which is found adaptively by the search. Atkinson et al., 2010 provide a general survey of the FS, with discussion. We have also implemented *Least Trimmed Squares*, (Hampel, 1975, Rousseeuw, 1984), as well as *Soft trimming (downweighting)*. Specifically we include S and MM estimation.

3.2 Robust Outlier Detection

Our algorithm works with k observations treated as outliers, providing the subset S_m of $m = n - k$ observations used in model fitting and parameter estimation. This section describes our outlier detection methods.

The default setting of the forward search uses the multivariate procedure of Riani et al. (2009) adapted for regression (Torti et al. 2021) to detect outliers at a simultaneous level of approximately 1% for samples of size up to around 1,000. Optionally, a different level can be selected. For the other two methods of robust regression, we apply a Bonferroni inequality to robust residuals to give a simultaneous test for outliers.

Since different response transformations can indicate different observations as outliers, the identification of outliers occurs repeatedly during our robust algorithm, once per iteration of the outer loop.

4 Improvements and Options

Our RAVAS procedure introduces five improvements to AVAS, programmed as options. These do not have a hierarchical structure, so that there are 2^5 possible choices of the options. The augmented star plot of Sect. 6 provides a method for assessing these choices. We discuss the motivation and implementation for each. The order in Sect. 4.1 is that in which the options are applied to the data when all five are used. We also give the names of the options, which are used as labels in the augmented star plot.

4.1 *Initial Calculations*

The structure of our algorithm is an elaboration of that of AVAS outlined in Sect. 2.3. Four of the five options can be invoked before the start of the outer loop.

4.1.1 Initialization of Data: Option Tyinitial

Our numerical experience is that it is often beneficial to start from a parametric transformation of the response. This is optionally found using the automatic robust procedure for power transformations described by Riani et al. (2022). For $\min(y) > 0$ we use the Box-Cox transformation. For $\min(y) \leq 0$ the extended Yeo-Johnson transformation is used (Atkinson et al. 2020). This family of transformations has separate Box and Cox transformations for positive and negative observations. In both cases the initial parametric transformations are only useful approximations, found by searching over a coarse grid of parameter values. The final non-parametric transformations sometimes suggest a generalization of the parametric ones.

4.1.2 Ordering Explanatory Variables in Backfitting: Option Scail

To avoid dependence of the fitted model on the order of the explanatory variables, one approach is to use an initial regression to remove the effect through scaling (Breiman 1988). With b_j the coefficient of $f_j(x)$ in the multiple regression of $g(y)$ on all $f_j(x)$, the option scail provides new transformed values for the explanatory variables: $\widehat{tX}_j = b_j f_j(x), j = 1, \ldots, p$. Option scail is used only in the initialization of the data.

4.1.3 Robust Regression and Robust Outlier Detection: Option Rob

We robustify our method through the use of robust regression as described in Sect. 3. The subset S_m, changing at each iteration, defines the observations used in backfitting and in the calculation of the variance stabilizing transformation.

4.1.4 Ordering Predictor Variables: Option OrderR2

For complete elimination of dependence on the order of the variables, we include an option that, at each iteration, provides an ordering which is based on the variable which produces the highest increment of R^2. With this option the most relevant features are immediately transformed and those that are perhaps irrelevant will be transformed in the final positions. For robust estimation, this procedure is applied solely to the observations in the subset S_m. Option orderR2 is available at each call to the backfitting function.

4.2 *Outer Loop*

4.2.1 Numerical Variance Stabilizing Transformation: Option Trapezoid

Plots of residuals against fitted values are widely used in regression analysis to check the assumption of constant variance. Here the observations have been transformed, so the fitted values are $\widehat{ty}_i$. To estimate the variance stabilizing transformation, the fitted values have to be sorted, giving a vector of ordered values $\widehat{ty}^s$. The residuals are ordered in the same way and, following the procedure of Sect. 2.4, provide estimates v_i of the integrand $V^{-0.5}(y)$ in (4). The v_i provide estimates at the ordered points $\widehat{ty_i}^s$. Calculation of the variance transformation (4) is however for sorted observed responses ty_i^s, rather than fitted, transformed responses $\widehat{ty_i}^s$. As did Tibshirani, we use the trapezoidal rule to approximate the integral. Linear interpolation and extrapolation are used in calculation of the v_i at the ty_i^s. We provide an option 'trapezoid' for the choice between two methods for the extrapolation of the variance function estimate, the interpolation method remaining unchanged. Our approach leads to trapezoidal summands in the approximation to the integral for the extrapolated elements, whereas Tibshirani's leads to rectangular elements. When we are concerned with robust inference, there are only $m = n - k$ members of $\widehat{ty}^s$ whereas there are n values of ty_i^s, so that robustness increases the effect of the difference between the two rules. The option trapezoid = false uses rectangular elements in extrapolation.

5 Simulations

We now use simulations to compare overall properties of AVAS with our robust version. The model was linear regression with data generated to have an average value of R^2 of 0.8. The responses were standardized to have zero mean and unit variance; 10% of the observations were contaminated by a shift δ and the responses were exponentiated. There were 1,000 simulations for $n = 200$ and $n = 1{,}000$ and 200 for $n = 10{,}000$. We encountered no numerical problems in the simulations. The figures compare the performances of RAVAS (with all options) and standard AVAS (with no options). Results for RAVAS use a dashed line.

The left-hand panels of Fig. 1 show the mean squared error of the parameter estimates in the linear models. For RAVAS, those for $n = 200$ and 1,000 exhibit a slight increase for moderately small values of δ which then decreases to be close to zero as δ increases and the outliers become easier to detect. That for $n = 10,000$ is virtually constant. The results for AVAS rapidly become much larger. The right-hand column shows the average power, that is the proportion of generated outliers that are detected by RAVAS. This climbs, in all cases, steadily to one. Of course, AVAS does not detect outliers.

We also compared the number of iterations to convergence of the algorithms; the default maximum is 20. Figure 2 shows results for the same simulations as above. The three panels show that RAVAS converges in around 3 iterations, except for $n = 200$ when there is a peak around $\delta = 2$, that is when the outliers are large enough to have

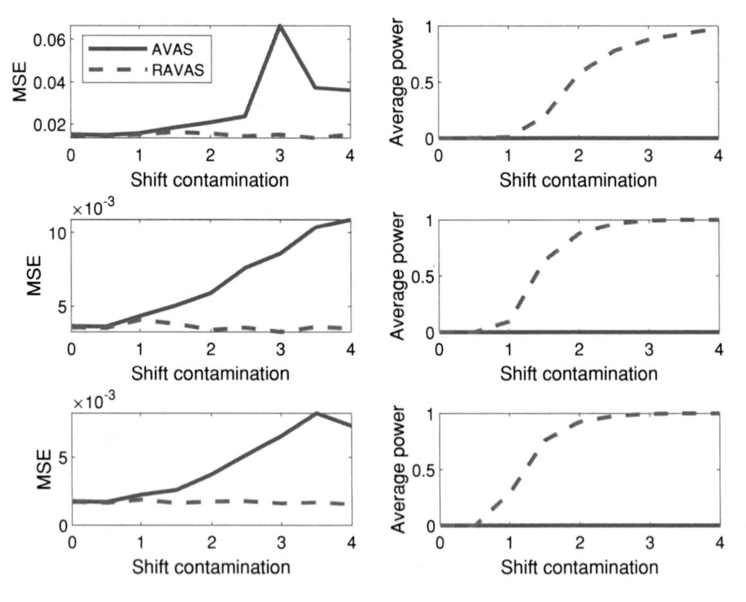

Fig. 1 Mean squared error (MSE) and average power. Top panels $n = 200$, $p = 5$, mid panels $n = 1{,}000$, $p = 10$, bottom panels $n = 10{,}000$, $p = 20$

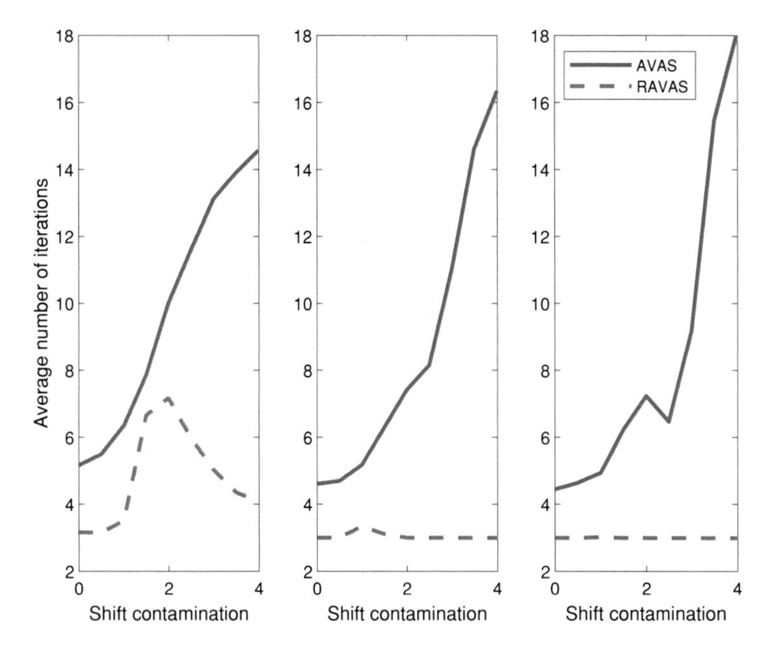

Fig. 2 Average number of iterations to convergence. Left-hand panel $n = 200$, $p = 5$, central panel $n = 1{,}000$, $p = 10$, right-hand panel $n = 10{,}000$, $p = 20$

an effect, but are still difficult to detect. This behavior is distinct from that of AVAS, where the number of iterations increases steadily both with δ and with the sample size.

6 The Generalized Star Plot

We have added five options to the original AVAS. There are therefore 32 combinations of options that could be chosen. It is not obvious that all will be necessary when analyzing any particular set of data. Our program provides flexibility in the assessment of these options. One possibility is a list of options ordered by, for example, the value of R^2 or of the significance of the Durbin-Watson test. In this section we describe the augmented star plot, one graphical method for visualizing interesting combinations of options in a particular data analysis. An example is Fig. 3.

We remove all analyses for which the residuals fail the Durbin-Watson test of independence and the Jarque-Bera normality test (Jarque and Bera 1987), at the 10 per cent level (two-sided for Durbin-Watson). The threshold of 10% can be optionally changed. We order the remaining, admissible, solutions by the Durbin-Watson significance level multiplied by the value of R^2 and by the number of units not declared as outliers. Other options are available. The lengths of rays in individual panels of

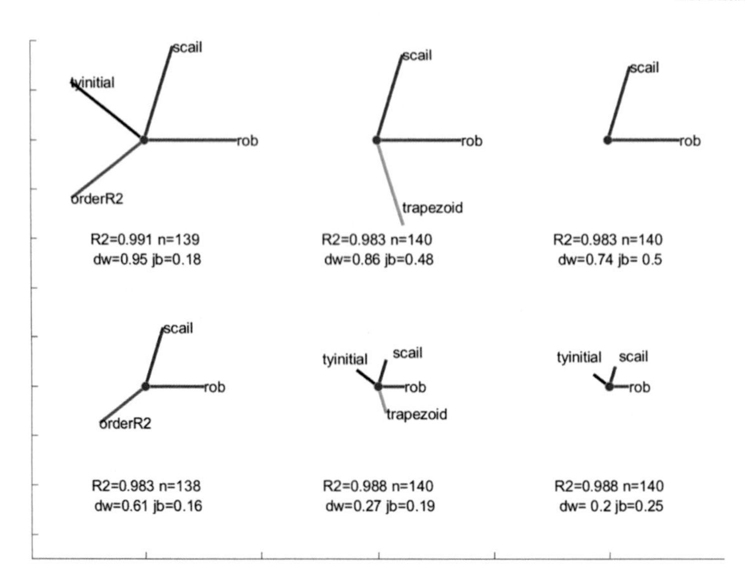

Fig. 3 Weight of fish. Augmented star plot of six options. Option 1 excludes trapezoid

the plot are of equal length for those features used in an analysis. All rays are in identical places in each panel of the plot; the length of the rays for each analysis are proportional to p_{DW}, the significance level of the Durbin-Watson test.

The ordering in which the five options are displayed in the plot depends on the frequency of their presence in the set of admissible solutions. For example, if robustness is the one which has the highest frequency, its ray is shown on the right. The remaining options are displayed counterclockwise, in order of frequency.

7 Prediction of the Weight of Fish

There are two websites, https://www.kaggle.com/aungpyaeap/fish-market and http://jse.amstat.org/datasets/fishcatch.txt which present data on the weight of 159 fish caught in a lake near Tampere, Finland. Interest is in the relationship between weight and five measurements of dimensions of the fish. There are 7 species of fish including pike. These behave rather differently from the other six species so we ignore them. We use the first three lengths for which the remaining fish seem homogenous. This assumption will be tested by our robust analysis if one or more species are identified as outliers. The variables are

y Weight of the fish (in grams)
x_1 Length from the nose to the beginning of the tail (in cm)
x_2 Length from the nose to the notch of the tail (in cm)
x_3 Length from the nose to the end of the tail (in cm).

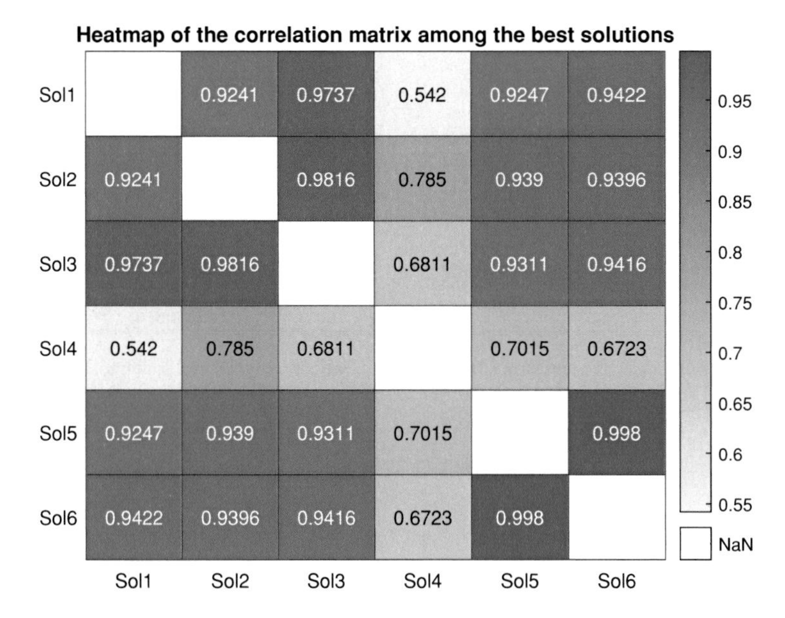

Fig. 4 Weight of fish. Heatmap of pairwise response correlations among the six solutions

After the deletion of the data on pike, 142 observations remain. Scatter plots of the response against the three explanatory variables reveal that all three lengths are highly correlated with the response, as they are with each other. It is reasonable to assume that weight increases with each of the explanatory variables. We therefore impose a monotonicity constraint on the transformations of the x_j. However, multiple regression with highly correlated explanatory variables can lead to problems in interpretation, such as estimated effects having a physically incorrect sign.

The augmented star plot for these data is in Fig. 3. There are six combinations of options that satisfy the constraints on the distribution of residuals. The first solution with an R^2 of 0.991 uses all five options except trapezoid. Robustness is used in all, succeeding selections giving R^2 values of 0.988 or 0.983.

The heatmap of the response correlations between the pairs of solutions is in Fig. 4. This shows that the first three solutions are strongly correlated with each other, as they are with the fifth and sixth solutions, the fifth and sixth solutions themselves having a very high correlation of 0.998. The heatmap emphasizes that solution four is appreciably different from the other five.

We now consider the adaptive identification of outliers using the FS. The first solution identifies three outliers. The left-hand panel of Fig. 5 shows that the response has been smoothly transformed. The plot of residuals against fitted values in the right-hand panel shows that there is only one remote outlier and that there is no remaining structure in the residuals. The plots of transformed explanatory variables (not given here) show that $f(x_1)$ is decreasing and slightly curved. The other two functions are increasing but only that for x_2 is almost straight, with slight curvature for the lowest values of the variable.

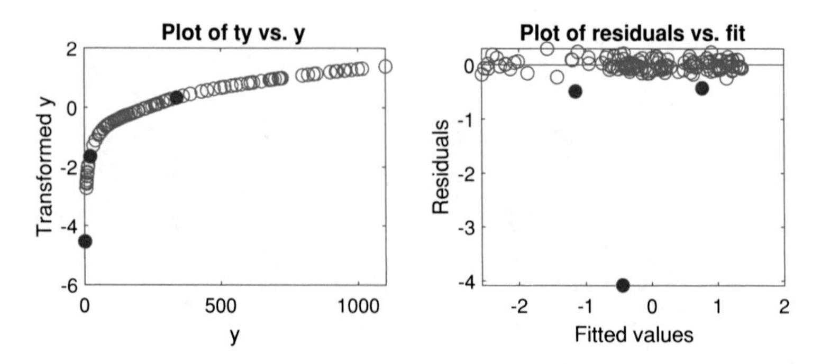

Fig. 5 Weight of fish. Left-hand panel, transformed y against y; right-hand panel, residuals against fitted values. Three outliers in red in the online version

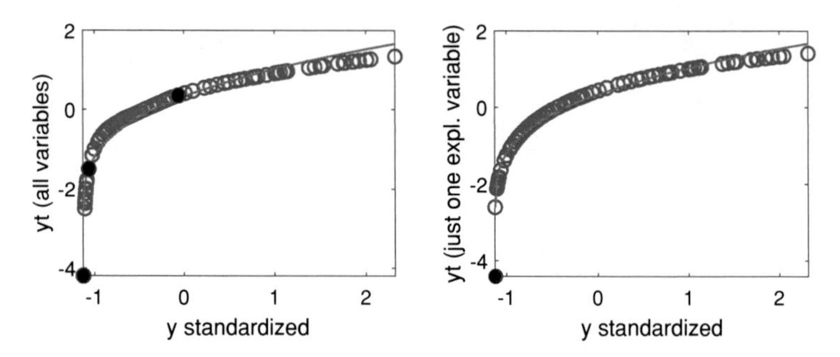

Fig. 6 Weight of fish. Non-parametric transformation of response y compared to $y^{1/3}$. Left-hand panel, three explanatory variables: right-hand panel, only x_1. Three and one outliers in black and red in the online version

The interpretation of the results from fitting three explanatory variables is that the variables are too highly correlated to give individually meaningful results. In our final analysis of the data we used only x_1. The star plot showed that the best selection included all options, except orderR2, which option is not possible with a single explanatory variable. The value of R^2 for this fit is 0.980 with the deletion of a single outlier. The three acceptable solutions had mutual fitted response correlations of 0.9994 or 1—the fitted model was stable to the choice of options.

In regressing volume on measurements of length, arguments from dimensional analysis suggest that volume should have a one-third transformation. Our final plot, Fig. 6, compares the transformed responses from the fits with three and one explanatory variables to $y^{1/3}$, for which transformation the value of $R^2 = 0.968$. The figure shows that both non-parametric transformations are indeed close to $y^{1/3}$ with a small systematic departure for the largest values of x. The fitted values from the single explanatory variable follow the power transformation slightly more closely than that when three variables are used. The transformation of x_1 is virtually straight with some curvature for large values. The flexibility of the non-parametric transformation provides an improved simple model compared with regression on untransformed x_1.

Acknowledgements We are very grateful to the editors and referees, whose comments greatly helped us to clarify the presentation of our work. Our research has benefited from the High Performance Computing (HPC) facility of the University of Parma. We acknowledge financial support from the University of Parma project "Robust statistical methods for the detection of frauds and anomalies in complex and heterogeneous data," and the Project ECS00000033 "Ecosystem for Sustainable Transition in Emilia-Romagna".

References

Atkinson, A. C., Riani, M., & Cerioli, A. (2010). The forward search: theory and data analysis (with discussion). *Journal of the Korean Statistical Society, 39*, 117–134. https://doi.org/10.1016/j.jkss.2010.02.007

Atkinson, A. C., Riani, M., & Corbellini, A. (2020). The analysis of transformations for profit-and-loss data. *Applied Statistics, 69*, 251–275. https://doi.org/10.1111/rssc.12389

Barlow, R. E., Bartholomew, D. J., Bremner, J. M., & Brunk, H. D. (1972). *Statistical inference under order restrictions*. Chichester: Wiley.

Box, G. E. P., & Cox, D. R. (1964). An analysis of transformations (with discussion). *Journal of the Royal Statistical Society, Series B, 26*, 211–252.

Box, G. E. P., & Tidwell, P. W. (1962). Transformations of the independent variables. *Technometrics, 4*, 531–550.

Breiman, L. (1988). Comment on "Monotone regression splines in action" (Ramsey, 1988). *Statistical Science, 3*, 442–445.

Buja, A., Hastie, T., & Tibshirani, R. (1989). Linear smoothers and additive models. *Annals of Statistics, 17*, 453–510.

Friedman, J., & Stuetzle, W. (1982). Smoothing of scatterplots. Technical report, Department of Statistics, Stanford University, Technical Report ORION 003.

Hampel, F. R. (1975). Beyond location parameters: robust concepts and methods. *Bulletin of the International Statistical Institute, 46*, 375–382.

Hastie, T., & Tibshirani, R. (1986). Generalized additive models. *Statistical Science, 1*, 297–318.

Hastie, T. J., & Tibshirani, R. J. (1990). *Generalized additive models*. London: Chapman and Hall.

Jarque, C. M., & Bera, A. K. (1987). A test for normality of observations and regression residuals. *International Statistical Review, 52*, 163–172.

Riani, M., Atkinson, A. C., & Cerioli, A. (2009). Finding an unknown number of multivariate outliers. *Journal of the Royal Statistical Society, Series B, 71*, 447–466.

Riani, M., Atkinson, A. C., & Corbellini, A. (2022). Automatic robust Box-Cox and extended Yeo-Johnson transformations in regression. *Statistical Methods and Applications*. https://doi.org/10.1007/s10260-022-00640-7.

Riani, M., Atkinson, A. C., & Corbellini, A. (2023). Robust transformations for multiple regression via additivity and variance stabilization. *Journal of Computational and Graphical Statistics*. https://doi.org/10.1080/10618600.2023.2205447.

Rousseeuw, P. J. (1984). Least median of squares regression. *Journal of the American Statistical Association, 79*, 871–880.

Tibshirani, R. (1988). Estimating transformations for regression via additivity and variance stabilization. *Journal of the American Statistical Association, 83*, 394–405.

Torti, F., Corbellini, A., & Atkinson, A. C. (2021). fsdaSAS: A package for robust regression for very large datasets including the Batch Forward Search. *Stats, 4*, 327–347.

A Random-Coefficients Analysis with a Multivariate Random-Coefficients Linear Model

Laura Marcis, Maria Chiara Pagliarella, and Renato Salvatore

Abstract Random-coefficients linear models can be considered as a particular case of linear mixed models. Different sources of variation are treated by random effects, which depend on some specific model design matrices. A redundancy analysis of estimates of the multivariate random effects may be able to capture the leading contribution to the covariance between the observed responses and the model covariates. We introduce the random effects of reduced space by a weighted least-squares closed-form solution, starting from the standardized multivariate best linear predictors. The application shows the effect of the linear dependence of the random effects in the space of the model covariates.

Keywords Random-coefficients model · Multivariate linear mixed model · Best linear unbiased predictor · Redundancy analysis

1 Introduction

In the standard linear regression models, a fundamental assumption is based on the independence of observations. This condition may be relaxed when clusters of units need to be explained following a common behavior inside a linear relationship. Usually, in the context of the mixed linear model (Demidenko 2004), it is possible to manage studies involving subjects naturally bound together. Widespread employment of general linear models with clustered data is common, for example, for longitudinal model analysis (Hsiao and Pesaran 2008).

L. Marcis · M. C. Pagliarella · R. Salvatore (✉)
University of Cassino and Southern Lazio, Cassino, Italy
e-mail: rsalvatore@unicas.it

L. Marcis
e-mail: laura.marcis@unicas.it

M. C. Pagliarella
e-mail: mc.pagliarella@unicas.it

L. Grilli et al. (eds.), *Statistical Models and Methods for Data Science*, Studies in Classification, Data Analysis, and Knowledge Organization,
https://doi.org/10.1007/978-3-031-30164-3_13

Random-coefficients linear regression models (RCM) (Longford 2011) represent a special class of linear mixed models, where the vector of regression coefficients for the subjects (e.g., repeated observations) is modeled in a second stage linear regression equation. In order to specify this type of models, it is convenient to define a two-stage hierarchical linear model, with a first stage that models within-subject observations, and a second stage with a linear model for the random regression coefficients.

Unlike the basic mixed model—in which random effects are not correlated with the response variable at the population level—in RCM this correlation depends on the fixed-effects design matrix of the regression model. Therefore, in RCM the correlation structure is more complicated (with respect to the simpler random intercept model) because it depends on the covariate values.
Since this happens, this paper aims to understand which random effect is most affected by the presence of the covariates and, at the same time, which random effect is "essentially random", i.e., orthogonal to the subspace spanned by the model covariates. We propose to achieve this result using an approach based on Redundancy Analysis (RDA) (Van Den Wollenberg 1977), for the fact that RDA provides a constrained analysis of the whole linear relations between the two sets of variables (Takane and Hwang 2007), and another unconstrained analysis given by the set of multivariate regression residuals (Härdle and Simar 2015). Some studies highlight that an RDA of the predicted criterion variables by the best linear unbiased predictor may be quite representative (Marcis and Salvatore 2020). This paper uses a RDA by a least-squares solution from the data provided by the random-coefficients predictors of the criterion variables. The application study performs the method introduced with the official data by the Italian Equitable and Sustainable Well-being indicators.

2 The Model and the Analysis of the Random Coefficients

Given a q-variate random vector $\mathbf{y}$, consider the case when a matrix $\mathbf{Y}$ of repeated observations from $\mathbf{y}$ is partitioned in m subjects (groups), each of them with n_i individuals ($i = 1, ..., m, ; j = 1, ..., n_i$). We assume that the general population model for the m subjects is $\mathbf{y}_i = \mathbf{B}'\mathbf{x}_i + \mathbf{A}_i\mathbf{z}_i$, where $\mathbf{B}$ is the $q \times p$ matrix of fixed regression coefficients. The $\mathbf{A}_i$ is a $q \times r$ matrix of q-variate r -dimensional vectors of random-effects, with $\mathbf{a}_i = \text{vec}(\mathbf{A}_i') \sim N(\mathbf{0}, \Sigma_a)$, $\Sigma_a = \text{cov}(\text{vec}(\mathbf{A}_i'))$, whose elements are $\{\Sigma_{a,qq'}\}$, with $\Sigma_{a,qq'} = \text{cov}(\text{vec}(\mathbf{A}'_{i,qq'}))$, the $r \times r$ blocks of Σ_a. When $r = p$, the population model is a multivariate RCM, with $\mathbf{z}_i = \mathbf{x}_i$. Given a sample of $N = \Sigma_i \Sigma_j n_{ij}$ units (e.g., repeated measurements), then the model structure is (with matrix dimensions reported as subscript):

$$\mathbf{Y}_{N\times q} = \mathbf{X}^{\otimes}_{N\times p}\mathbf{B}_{p\times q} + \mathbf{Z}^{\otimes}_{N\times pm}\mathbf{A}_{pm\times q} + \mathbf{E}_{N\times q}, \tag{1}$$

with $\mathbf{X}^{\otimes}$ the matrix of data covariates, $\mathbf{Z}^{\otimes}$ the design matrix of random effects (the exponent $\otimes$, the symbol usually used for the Kronecker product, is to differentiate the matrices just introduced from those that will be introduced below in order to rewrite the model in vector form) and $\mathbf{E}$ the matrix of regression within-subject errors, with $\text{vec}(\mathbf{E}) \sim N(\mathbf{0}, \boldsymbol{R})$. In this article, we assume both $\mathbf{Y}$ and $\mathbf{X}^{\otimes}$ as columnwise centered and standardized. With $\otimes$ the Kronecker product and $\text{cov}(\mathbf{y}^*)$ the model covariance, we rewrite the last model in the vector form, with $\mathbf{y}^* = \text{vec}(\mathbf{Y})$, $\mathbf{X} = (\mathbf{I} \otimes \mathbf{X}^{\otimes})$, $\boldsymbol{\beta} = \text{vec}(\mathbf{B})$, $\mathbf{Za} = \text{vec}(\mathbf{ZA}) = (\mathbf{I} \otimes \mathbf{Z}^{\otimes})\text{vec}(\mathbf{A})$.

The components of the model covariance $\Sigma_a = \Sigma_a(\boldsymbol{\theta})$ depend on a multivariate vector $\boldsymbol{\theta}$ that accounts for some variance-covariance parameters, to be estimated. Maximum and restricted maximum likelihood procedures may be employed for the estimation in case of normality, as well as other general methods, like method of moments. Some studies report the effectiveness of the method of moments as the most common restricted likelihood estimation (see again Demidenko 2004), that's the one used in this work. By the following Theorem 1 and Proposition 1, we extend some properties of the univariate RCM to the multivariate RCM (the proofs are given in the Appendix).

Theorem 1 *Conditional distribution of the multivariate random effects*

Given the multivariate RCM in (1)*, with*

$$\mathbf{a} \sim N(\mathbf{0}, \boldsymbol{D}), \quad E(\mathbf{y}^*|\mathbf{a}) \sim N(\mathbf{X}\boldsymbol{\beta}, \mathbf{ZDZ}' + \mathbf{R}), \quad \boldsymbol{\alpha} = (\mathbf{a}', \mathbf{y}^{*'})',$$
$$\text{cov}(\boldsymbol{\alpha}) = \Omega = \begin{bmatrix} \Omega_{a,a} & \Omega_{a,y^*} \\ \Omega_{y^*,a} & \Omega_{y^*,y^*} \end{bmatrix}, \textit{ and}$$
$$\text{cov}(\mathbf{a}, \mathbf{y}^*) = \Omega_{\mathbf{a},\mathbf{y}^*} = \Omega'_{\mathbf{y}^*,\mathbf{a}} = \mathbf{D}[\mathbf{I} \otimes (\mathbf{Z}^{\otimes})'],$$

then the conditional distribution of $\mathbf{a}$ *given* $\mathbf{y}^*$ is

$$\mathbf{a}|\mathbf{y}^* \sim N[\Omega_{\mathbf{a},\mathbf{y}^*}\Omega^{-1}_{\mathbf{y}^*,\mathbf{y}^*}(\mathbf{y}^* - \mathbf{X}\boldsymbol{\beta}), \Omega_{\mathbf{a},\mathbf{a}} - \Omega_{\mathbf{a},\mathbf{y}^*}\Omega^{-1}_{\mathbf{y}^*,\mathbf{y}^*}\Omega_{\mathbf{y}^*,\mathbf{a}}].$$

Proposition 1 *Component estimates of the multivariate RCM*

Given the multivariate RCM in (1)*, with*

$$\mathbf{a} \sim N(\mathbf{0}, \mathbf{D}),$$
$$E(\mathbf{y}^*|\mathbf{a}) \sim N(\mathbf{X}\boldsymbol{\beta}, \mathbf{ZDZ}' + \mathbf{R}),$$
$$\mathbf{Z}^{\otimes} = \mathbf{I} \otimes \mathbf{X}_d, \quad \mathbf{X}_d = \text{diag}(\mathbf{X}_1, \ldots, \mathbf{X}_m), \quad \text{cov}(\text{vec}(\mathbf{A}_i')) = \Sigma_{\mathbf{a}}(\boldsymbol{\theta}),$$

and an estimate $\widehat{\boldsymbol{\theta}}$*, the model estimates of* $\widehat{\boldsymbol{\beta}} - \boldsymbol{\beta}$ and $\mathbf{Z}(\widetilde{\mathbf{a}} - \mathbf{a})$ *have the covariance matrix* Ψ*, with elements*

$$\begin{aligned}
\Psi_{11} &= \text{cov}(\widehat{\boldsymbol{\beta}} - \boldsymbol{\beta}) = \text{cov}(\widehat{\boldsymbol{\beta}}) = [\mathbf{X}'(\mathbf{ZDZ}' + \mathbf{R})^{-1}\mathbf{X}]^{-1},\\
\Psi_{12} &= \Psi_{21}' = \text{cov}[\widehat{\boldsymbol{\beta}}, \mathbf{Z}(\widetilde{\mathbf{a}} - \mathbf{a})]\\
&= -[\mathbf{X}'\text{cov}(\mathbf{y}^*)^{-1}\mathbf{X}]^{-1}\mathbf{X}'\text{cov}(\mathbf{y}^*)^{-1}(\mathbf{I} \otimes \mathbf{X}_d)\mathbf{D}(\mathbf{I} \otimes \mathbf{X}_d)',\\
\Psi_{22} &= \text{cov}[\mathbf{Z}(\widetilde{\mathbf{a}} - \mathbf{a})] = \mathbf{ZDZ}' - \mathbf{ZDZ}'\mathbf{PZDZ}'\\
&= (\mathbf{I} \otimes \mathbf{X}_d)\mathbf{D}(\mathbf{I} \otimes \mathbf{X}_d') - (\mathbf{I} \otimes \mathbf{X}_d)\mathbf{D}(\mathbf{I} \otimes \mathbf{X}_d') \times \mathbf{P} \times (\mathbf{I} \otimes \mathbf{X}_d)\mathbf{D}(\mathbf{I} \otimes \mathbf{X}_d').
\end{aligned}$$

The matrix $\mathbf{P}$ *represents the projection matrix onto the complement of the column space of* $\mathbf{X}$ *in the metrics of* $(\mathbf{ZDZ}' + \mathbf{R})^{-1}$.

The multivariate linear best predictor $\widetilde{\mathbf{Y}}$ is given by the best linear predictor $\widetilde{\mathbf{y}}^* = \text{vec}(\widetilde{\mathbf{Y}}) = \mathbf{X}\widehat{\boldsymbol{\beta}} + \mathbf{Z}\widetilde{\mathbf{a}}$, $\widetilde{\boldsymbol{a}} = E(\boldsymbol{a}|\boldsymbol{y})$. Here

$$\begin{aligned}
&\widehat{\boldsymbol{\beta}} = \widehat{\boldsymbol{\beta}}_{gls}, \mathbf{Z}\widetilde{\mathbf{a}} = \mathbf{ZDZ}'\text{cov}(\mathbf{y}^*)^{-1}(\mathbf{y}^* - \mathbf{X}\widehat{\boldsymbol{\beta}}),\\
&\widetilde{\mathbf{y}}^* = \Gamma\mathbf{y}^* + (\mathbf{I} - \Gamma)\mathbf{X}\widehat{\boldsymbol{\beta}},\\
&\Gamma = \mathbf{ZDZ}'\text{cov}(\mathbf{y}^*)^{-1}, \text{cov}(\mathbf{y}^*) = \mathbf{ZDZ}' + \mathbf{R}, \text{ and}\\
&\mathbf{D} = \left\{\mathbf{D}_{qq'}\right\}, \text{with}\mathbf{D}_{qq'} = \Sigma_{\mathbf{a},\mathbf{qq}'} \otimes \mathbf{I}_m.
\end{aligned}$$

Given $\widetilde{\mathbf{y}}_{qi}$, for the RCM as a special case of the general linear mixed model, we have that $\text{cov}(\mathbf{a}_{qi}, \mathbf{y}_{qi}) = \mathbf{DZ}_i' = \mathbf{DX}_{qi}'$. Thus, for the RCM in (1) $\text{cov}(\mathbf{a}, \mathbf{y}^*) = \mathbf{D}(\mathbf{I} \otimes \mathbf{X}_d')$ (with $r = p$):

$$\begin{aligned}
\text{cov}(\mathbf{a}, \mathbf{y}^*) &= E(\mathbf{a}, \mathbf{y}^{*'}) = \text{cov}(\mathbf{a}, \mathbf{a}'\mathbf{Z}') = \mathbf{DZ}' = \mathbf{D}(\mathbf{I} \otimes \mathbf{Z}^{\otimes})'\\
&= \mathbf{D}[\mathbf{I} \otimes (\mathbf{Z}^{\otimes})'] = \mathbf{D}(\mathbf{I} \otimes \mathbf{X}_d').
\end{aligned}$$

For sake of clarity, it should be noted that $\widetilde{\mathbf{a}}$ depend on the matrix of fixed effects $\mathbf{X}^{\otimes}$ and that, unlike the classic mixed model in which $\text{cov}(\boldsymbol{a}, \boldsymbol{y}^*) = \boldsymbol{DZ}'$, in this case $\text{cov}(\mathbf{a}, \mathbf{y}^*) = \mathbf{D}(\mathbf{I} \otimes \mathbf{X}_d')$ that is, the random effects are affected by the covariates. In a similar way to RDA, we can achieve our goal by evaluating which random effects are more influenced by the model covariates (and, as a consequence, which random effects are not).

In standard RDA two subspaces are obtained: one constrained—from the Singular Value Decomposition (SVD) of the predicted values, $\widehat{\mathbf{Y}} = \mathbf{X}\widehat{\mathbf{B}} = \mathbf{U}_{\widehat{\mathbf{Y}}}\Lambda\mathbf{V}_{\widehat{\mathbf{Y}}}'$—and another unconstrained—from the SVD of the residual values.

Our new step is as follows: after estimating the vector of the multivariate covariance parameters $\boldsymbol{\theta} = \widehat{\boldsymbol{\theta}}$, and the model estimates of fixed and random effects $\widetilde{\mathbf{Y}} = \mathbf{X}^{\otimes}\widehat{\mathbf{B}}(\widehat{\theta}) + \mathbf{Z}^{\otimes}\widetilde{\mathbf{A}}(\widehat{\theta})$, we want to use the projection of the centered and standardized ratio $\boldsymbol{\phi} = \text{cov}(\widetilde{\mathbf{y}})^{-\frac{1}{2}}[\widetilde{\mathbf{y}} - E(\widetilde{\mathbf{y}})] = \text{cov}(\widetilde{\mathbf{y}})^{-\frac{1}{2}}\widetilde{\mathbf{y}}^{**}$ of the multivariate predictor $\widetilde{\mathbf{y}}^*$, in a common reduced covariate subspace.

In particular, assuming both $\mathbf{Y}$ and $\mathbf{X}^{\otimes}$ as columnwise centered and standardized, and a balanced design, we get in the general RCM $\text{cov}(\mathbf{a}_{qi}, \mathbf{y}_{qi}) = \mathbf{DZ}_i' = \mathbf{DX}_{qi}'$. Then, we can compute $\boldsymbol{\phi}$ as indicated above, noting that $E(\widetilde{\mathbf{y}}) = E(\mathbf{y}) = \mathbf{X}\beta$, $\widetilde{\mathbf{y}} - E(\widetilde{\mathbf{y}}) = \widetilde{\mathbf{y}}^* - \mathbf{X}\widehat{\boldsymbol{\beta}} = \mathbf{Z}\widetilde{\mathbf{a}} - E(\mathbf{Z}\widetilde{\mathbf{a}}) = \mathbf{Z}\widetilde{\mathbf{a}}$.

Therefore, $\mathrm{cov}(\widetilde{\mathbf{y}}^*)^{-\frac{1}{2}}\mathbf{Z}\widetilde{\mathbf{a}} = \mathrm{cov}(\widetilde{\mathbf{y}})^{-\frac{1}{2}}\widetilde{\mathbf{y}}^{**} = \boldsymbol{\phi}$. In certain cases, due to the unbiasedness of model covariance parameters estimates, the expression $\mathrm{cov}(\widetilde{\mathbf{y}}^*)$ may be substituted by the mean squared error $\mathrm{mse}(\widetilde{\mathbf{y}}^*)$. Consequently, $\mathrm{mse}(\widetilde{\mathbf{y}}^*)^{-\frac{1}{2}}$ represents the inverse matrix of the root mean squared error of the multivariate predictor.

We may adopt two different methods to achieve a numerical evaluation of $\mathrm{cov}(\widetilde{\mathbf{y}})^{-\frac{1}{2}}\widetilde{\mathbf{y}}^{**}$, or simplifying, the statistic $\boldsymbol{\phi} = \mathrm{cov}(\widetilde{\mathbf{y}}^*)^{-\frac{1}{2}}\mathbf{Z}\widetilde{\mathbf{a}}$:

(a) <u>*First method:*</u> compute $\mathbf{M} = \mathrm{vecdiag}\left\{\mathrm{cov}(\widetilde{\mathbf{y}}^*)^{-1}\right\} = \mathbf{I}_N \circ \mathrm{cov}(\widetilde{\mathbf{y}}^*)^{-1}$, as the matrix of diagonal elements of $\mathrm{cov}(\widetilde{\mathbf{y}}^*)^{-\frac{1}{2}}$ ($\circ$ is the Hadamard product). Then compute $\mathbf{H} = [\mathbf{I}_N(\mathbf{1}'_N \otimes \mathbf{Z}\widetilde{\mathbf{a}})]$, with $\mathbf{Z}\widetilde{\mathbf{a}}$ as the vector as diagonal elements of the diagonal matrix $\mathbf{H}$, and (by the vec^{-1} operator), we get $\mathbf{Z}\widetilde{\mathbf{a}} \times cov(\widetilde{\mathbf{y}}^*)^{-\frac{1}{2}} = \mathbf{H}\mathbf{M}^{-\frac{1}{2}}$, and $\widetilde{\widetilde{\mathbf{y}}}^* = \mathbf{H}\mathbf{M}^{-\frac{1}{2}}\mathbf{1}_N$.
Finally, $\widetilde{\widetilde{\mathbf{Y}}} = vec^{-1}(\widetilde{\widetilde{\mathbf{y}}}^*) = ((\mathrm{vec}\mathbf{I}_q)' \otimes \mathbf{I}_N)(\mathbf{I}_q \otimes \widetilde{\widetilde{\mathbf{y}}}^*) = ((\mathrm{vec}\mathbf{I}_q)' \otimes \mathbf{I}_N)(\mathbf{I}_q \otimes \mathbf{H}\mathbf{M}^{-1}\mathbf{1}_N)$.

(b) <u>*Second method:*</u> compute $\mathbf{M}^{-\frac{1}{2}} = \{\mathbf{I}_N \circ \mathrm{cov}[\widetilde{\mathbf{y}}^*]\}^{-\frac{1}{2}}$.
Then, with $\boldsymbol{\tau} = \{\mathbf{I}_N \circ \mathrm{cov}(\widetilde{\mathbf{y}}^*)\}^{-\frac{1}{2}}\mathbf{1}_N$ as the vector of the inverse diagonal elements of $\mathrm{cov}(\widetilde{\mathbf{y}}^*)$, and $\mathbf{T} = \mathrm{vec}^{-1}(\boldsymbol{\tau}) = ((\mathrm{vec}\mathbf{I}_q)' \otimes \mathbf{I}_N)(\mathbf{I}_q \otimes \boldsymbol{\tau})$, the matrix of elementwise variances of $\widetilde{\mathbf{Y}}$, we get $\widetilde{\widetilde{\mathbf{Y}}} = \widetilde{\mathbf{Y}} \circ \mathbf{T}$.

Since we are interested in the simultaneous representation of all the predicted $\widetilde{\mathbf{a}}_i$, given by a common projection subspace, we may also proceed following an alternative method, respect to those under methods (a) and (b), to explore the ratio $\boldsymbol{\phi}$.

(c) <u>*Third method:*</u> Find the minimum Frobenius norm from the multivariate predictor $\widetilde{\mathbf{Y}}$, as explained below.

The minimum Frobenius norm from the multivariate predictor $\widetilde{\mathbf{Y}}$ is found by the difference

$$\Phi = \widetilde{\mathbf{Y}}^{**}\mathrm{var}(\widetilde{\mathbf{y}})^{-\frac{1}{2}} - \mathbf{X}^{\otimes}\mathbf{B},\ \widetilde{\mathbf{Y}}^{**} = \widetilde{\mathbf{Y}} - E(\widetilde{\mathbf{Y}}) = \widetilde{\mathbf{Y}} - \mathbf{1}_N E(\widetilde{\mathbf{y}}'),\ \text{i.e.,}$$
$$\|\Phi\|^2 = trace(\Phi'\Phi) = \left\|\widetilde{\mathbf{Y}}^{**}\Sigma^{-\frac{1}{2}} - \mathbf{X}^{\otimes}\mathbf{B}\right\|^2 = \min .$$

We assume

$$\mathrm{cov}(\widetilde{\mathbf{y}}^*_{qq'}) = E\left\{(\widetilde{\mathbf{y}}^*_q - \mathbf{y}^*_q)(\widetilde{\mathbf{y}}^*_{q'} - \mathbf{y}^*_{q'})'\right\},$$
$$\Sigma = \frac{1}{N}\left\{\mathrm{gtrace}[\mathrm{cov}(\widetilde{\mathbf{y}}^*_{qq'})]\right\},\ \text{and}$$
$$\mathrm{gtrace}[\mathrm{cov}(\widetilde{\mathbf{y}}^*_{qq'})] = \mathrm{trace}[E\left\{(\widetilde{\mathbf{y}}^*_q - \mathbf{y}^*_q)(\widetilde{\mathbf{y}}^*_{q'} - \mathbf{y}^*_{q'})'\right\}] = E\left\{(\widetilde{\mathbf{y}}^*_{q'} - \mathbf{y}^*_{q'})'(\widetilde{\mathbf{y}}^*_q - \mathbf{y}^*_q)\right\}.$$

Here $\mathrm{gtrace}[\mathrm{cov}(\widetilde{\mathbf{y}}^*_{qq'})]$ is the generalized trace operator, that gives a matrix with elements as traces of submatrices of a given square matrix (Timm 2022). Now, setting

$$\boldsymbol{\phi} = \text{vec}(\Phi) = (\Sigma^{-\frac{1}{2}} \otimes \mathbf{I}_N)\widetilde{\mathbf{y}}^* - (\mathbf{I}_q \otimes \mathbf{X}^{\otimes})\overline{\boldsymbol{\beta}} = \overline{\Sigma}^{-\frac{1}{2}}\widetilde{\mathbf{y}}^* - \mathbf{X}\overline{\boldsymbol{\beta}},$$
$$\overline{\boldsymbol{\beta}} = \text{vec}(\mathbf{B}), \overline{\Sigma}^{-\frac{1}{2}} = (\Sigma^{-\frac{1}{2}} \otimes \mathbf{I}_N),$$

we come to the following properties of $\overline{\boldsymbol{\beta}}$:

$$\text{trace}(\boldsymbol{\phi}'\boldsymbol{\phi}) = \text{trace}\left\{(\overline{\Sigma}^{-\frac{1}{2}}\widetilde{\mathbf{y}}^* - \mathbf{X}\overline{\boldsymbol{\beta}})'(\overline{\Sigma}^{-\frac{1}{2}}\widetilde{\mathbf{y}}^* - \mathbf{X}\overline{\boldsymbol{\beta}})\right\}$$
$$= (\widetilde{\mathbf{y}}^* - \overline{\mathbf{X}\boldsymbol{\beta}})'\overline{\Sigma}^{-1}(\widetilde{\mathbf{y}}^* - \overline{\mathbf{X}\boldsymbol{\beta}}),$$

where $\overline{\mathbf{X}} = \overline{\Sigma}^{\frac{1}{2}}\mathbf{X}$. Thus, $\widehat{\overline{\boldsymbol{\beta}}} = (\overline{\mathbf{X}}'\overline{\Sigma}^{-1}\overline{\mathbf{X}})^{-1}\overline{\mathbf{X}}'\overline{\Sigma}^{-1}\widetilde{\mathbf{y}}^*$ is the q-variate vector in the subspace spanned by the columns of the matrix $\overline{\mathbf{X}}$, with $\widetilde{\mathbf{y}}^*$ orthogonal to the columns of $\overline{\mathbf{X}}$ in the metrics of $\overline{\Sigma}^{-1}$, $\widetilde{\mathbf{y}}^{*\prime}\overline{\Sigma}^{-1}\overline{\mathbf{x}} = 0$. Then, $\mathbf{P}_{\overline{\mathbf{X}}} = \overline{\mathbf{X}}(\overline{\mathbf{X}}'\overline{\Sigma}^{-1}\overline{\mathbf{X}})^{-1}\overline{\mathbf{X}}'\overline{\Sigma}^{-1}$ is the projection matrix of the predictor $\widetilde{\mathbf{y}}$ onto the joint subspace by $\overline{\mathbf{X}}$. The SVD of $\overline{\overline{\mathbf{Y}}} = \mathbf{X}^{\otimes}\widehat{\overline{\boldsymbol{\beta}}}$ gives the common rescaled predictor's coordinates of $\widetilde{\boldsymbol{Y}}$, i.e., $\overline{\overline{\boldsymbol{Y}}} = \boldsymbol{U}_{\overline{\overline{Y}}}\boldsymbol{\Lambda}_{\overline{\overline{Y}}}\boldsymbol{V}'_{\overline{\overline{Y}}}$, further noticing that $\boldsymbol{U}^{\star}_{\overline{\overline{Y}}} = \widetilde{\boldsymbol{Y}}\boldsymbol{V}_{\overline{\overline{Y}}}\boldsymbol{\Lambda}^{-1}_{\overline{\overline{Y}}}$ contains the row joint reduced coordinates of $\widetilde{\boldsymbol{Y}}$ in the space of $\overline{\overline{\boldsymbol{Y}}}$.

Finally, to give the exact number of the principal coordinates by the number of non-zero eigenvalues when projecting the matrix Φ in the two different subspaces, depending or not from the design matrix $\mathbf{X}$, we establish the following Theorem on the rank of Φ (the proof is given in the Appendix).

Theorem 2 *Rank of* Φ

Let Φ_i *a* $n_i \times p$ *matrix of* n_i *repeated observations from the* i^{th} *subject* ($i = 1, ..., m$, $\Sigma_{i=1}^m n_i = n$) *of the p-variate random vector* $\boldsymbol{\phi}_{ij}$, *where the* j^{th} *row vector* ($j = 1, ..., n_i$) *of* Φ_i *has the linear structure*

$$\boldsymbol{\phi}'_{ij} = \mathbf{B}'\mathbf{x}'_{ij} + \boldsymbol{\alpha}'_i + \boldsymbol{\eta}'_{ij},$$

where $\boldsymbol{\alpha}_i \sim \text{ind}(\mathbf{0}, \Sigma_{\boldsymbol{\alpha}})$, $\boldsymbol{\eta}_{ij} \sim \text{ind}(\mathbf{0}, \sigma^2)$, $\text{vec}(\boldsymbol{\eta}'_i) \sim (\mathbf{0}, \sigma^2 I_{n_i \times p})$, $\text{cov}(\boldsymbol{\alpha}_i, \boldsymbol{\eta}_i) = 0$. $\mathbf{B}$ *is a* $h \times p$ *matrix of fixed effects in the linear model, and* $\boldsymbol{\alpha}'_i$ *is a vector of p-dimensional random effects. Let also* $\Phi = [\Phi_1, \Phi_2, ..., \Phi_m]'$, *modeled as*

$$\Phi = \mathbf{X}\mathbf{B} + \mathbf{Z}\boldsymbol{\alpha} + \boldsymbol{\eta}, \quad \mathbf{Z} = \text{diag}(\mathbf{1}_{n_1}, ..., \mathbf{1}_{n_m}), \quad \text{cov}(\text{vec}(\boldsymbol{\alpha})) = \Sigma_{\boldsymbol{\alpha}} \otimes \mathbf{I}_m.$$

Then, for $n \gg p \gg h$, *and* $p > h + m$, *we have the following rank* $\mathbf{r}$ *of the model components:*

(a) $\mathbf{r}(\mathbf{XB}) = h$
(b) $\mathbf{r}(\mathbf{Z}\alpha) = m - 1$,
(c) $\mathbf{r}(\mathbf{XB}) + \mathbf{r}(\mathbf{Z}\alpha) = h + m - 1$,
(d) $\mathbf{r}(\Phi) = \mathbf{r}(\eta) = p$.

If $p \leq h + m$, *we have that:*

(e) $\mathbf{r}(\mathbf{XB}) + \mathbf{r}(\mathbf{Z}\alpha) = p$.

3 Application Study

In accordance with the recent law reforms in Italy, the Equitable and Sustainable Well-being indicators (in Italian, BES)—annually provided by the Italian Statistical Institute (ISTAT, 2017)—are designed to define the economic policies which largely act on some fundamental aspects of the quality of life. In order to highlight the result of the proposed method we use 12 BES indicators relating to the years 2013–2016, collected at NUTS-2 (Nomenclature of Territorial Units for Statistics-2) level. The variables employed in the application study are in Table 1.

We use the per capita adjusted disposable income variable (its logarithm, as is usually done in economics studies)—indicated with BE1—as an unique covariate for the RCM, while the remaining 11 variables are dependent variables (please, refer to Table 1 again for the description and acronyms used for the variables). The application uses the restricted maximum likelihood estimation and a Sas/IML code. We adopted the estimation method (a) discussed above. We propose to estimate the model under an uniform correlation structure among the multivariate components of the random effects. This structure is equivalent to the compound-symmetry covariance structure, with a better numerical property in terms of optimization. Further, some studies (Ledoit and Wolf 2004) highlight that using uniform correlation matrices reduces the estimation noise. The model covariance matrix $\mathbf{D}$ of random effects is then a

Table 1 Description of the variables used for the application

Variables	Description
S8	Age-standardized mortality rate for dementia and nervous system diseases
IF3	People having completed tertiary education (30–34 years old)
L12	Share of employed persons who feel satisfied with their work
REL4	Social participation
POL5	Trust in other institutions like the police and the fire brigade
SIC1	Homicide rate
BS3	Positive judgement for future perspectives
PATR9	Presence of Historic Parks/Gardens and other Urban Parks recognized of significant public interest
AMB9	Satisfaction for the environment—air, water, noise
INN1	Percentage of R&D expenditure on GDP
Q2	Children who benefited of early childhood services
BE1	Per capita adjusted disposable income
LBE1	Logarithm of Per capita adjusted disposable income

Table 2 The slope parameters by the multivariate regression with the LBE1 covariate

Dependent variable	Slope parameter (LBE1)	STD Error	t	Pr >t
AMB9	0.9802	0.3255	3.01	0.0035
BS3	0.9330	0.0891	10.47	0.0001
IF3	−0.3166	0.1673	−1.89	0.0621
INN1	−0.0433	0.0170	−2.54	0.0130
L12	0.0016	0.0107	0.15	0.8786
PATR9	0.0975	0.0756	1.29	0.2007
POL5	−0.0036	0.0085	−0.42	0.6775
Q2	0.2031	0.1762	1.15	0.2526
REL4	0.5602	0.1690	3.31	0.0014
S8	−0.0506	0.0293	−1.73	0.0879
SIC1	−0.0072	0.0150	−0.48	0.6314

generalized uniform correlation matrix, $\mathbf{D} = \Sigma_a \boxtimes \mathbf{I}_m$. The symbol $\boxtimes$ is the Tracy-Singh product operator that represents a Kronecker product structure applied on matrices instead of their elements (Ploymukda et al. 2018). The multivariate model is based on two covariance parameters, σ_a^2 and ρ_a, that represent jointly variances and covariances, and correlations within and between the random effects of the multivariate dependent model components. Analytically, the covariance matrices, whit the symbols defined above in the previous sections, are

$$\begin{aligned}
\Sigma_a &= \mathbf{I}_q \otimes \Sigma_{a,q} + (\mathbf{1}_q\mathbf{1}_q' - \mathbf{I}_q) \otimes \Sigma_{a,qq'} \\
&= \mathbf{I}_q \otimes (\Sigma_{a,q} - \Sigma_{a,qq'}) + (\mathbf{1}_q\mathbf{1}_q') \otimes \Sigma_{a,qq'} \\
\Sigma_{a,qq'} &= \sigma_a^2 \rho_a \mathbf{1}_r\mathbf{1}_r' \\
\Sigma_{a,q} &= \sigma_a^2[(1-\rho_a)\mathbf{I}_r + \rho_a \mathbf{1}_r\mathbf{1}_r'] = \sigma_a^2[(1-\rho_a)\mathbf{I}_r] + \Sigma_{a,qq'}
\end{aligned}$$

The restricted maximum likelihood estimates of the covariance parameters are $\widehat{\sigma}_a^2 = 7.315$ and $\widehat{\rho}_a = 0.083$. Table 2 shows the slope parameter estimates from the multivariate regression, with their significance level. Table 3 reports the MANOVA multivariate test statistics, based on the characteristic roots. These are the eigenvalues of the product of the sum-of-squares matrix of the regression model and the sum-of-squares matrix of the error. The null hypothesis for each of these tests is the same: the independent variable (LBE1) has no effect on any of the dependent variables. The four tests are all significant.

After the application of the multivariate model in (1), we may explore the global set of relations between the variables employed, by applying the redundancy analysis of Φ. Figures 1 and 2 give the two independent subspaces by the projection of the matrix Φ onto the space of the model (1), and of its orthogonal complement.

Table 3 MANOVA test criteria and F approximations for the hypothesis of no overall LBE1 effect

Statistic	Value	F value	Num DF	Den DF	Pr > F
Wilks lambda	0.0566	102.98	11	68	<0.0001
Pillai's trace	0.9434	102.98	11	68	<0.0001
Hotelling–Lawley trace	16.6590	102.98	11	68	<0.0001
Roy's largest root	16.6590	102.98	11	68	<0.0001

The first two eigenvalues of the constrained analysis in the space of the standardized predictor explain the 53.34 and 37.45 per cent of the total variability. The random effects of the model for variables IF3 (People having completed tertiary education (30–34 years old)) and Q2 (Children who benefited of early childhood services) are those most responsible in explaining the correlation between the multivariate response in $\boldsymbol{Y}$ and the LBE1 model covariate (logarithm of the individual income) (Fig. 1). This means that random clusters by the dependent variables IF3 and Q2 are those that explain at best the inter-regional variability expressed by the multivariate random effects.

Conversely, the analysis of the residuals with the unconstrained analysis reports (Fig. 2) variables AMB9 (Satisfaction for the environment—air, water, noise) and REL4 (Social participation) as the most uncorrelated with LBE1 through the model random effects. This means that differences between the regions (by the values of the random effects) for these variables tend to be independent of the individual

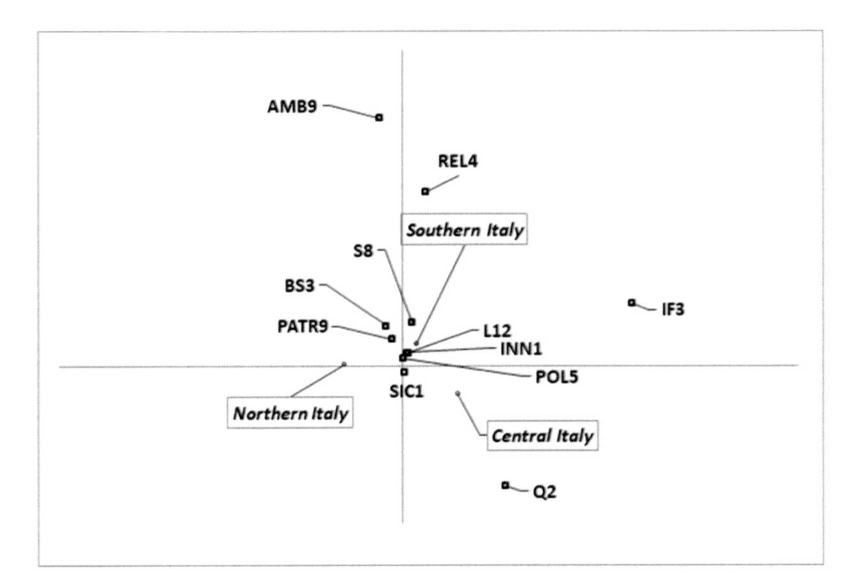

Fig. 1 Constrained analysis of the matrix Φ in the space of $\mathbf{X}$

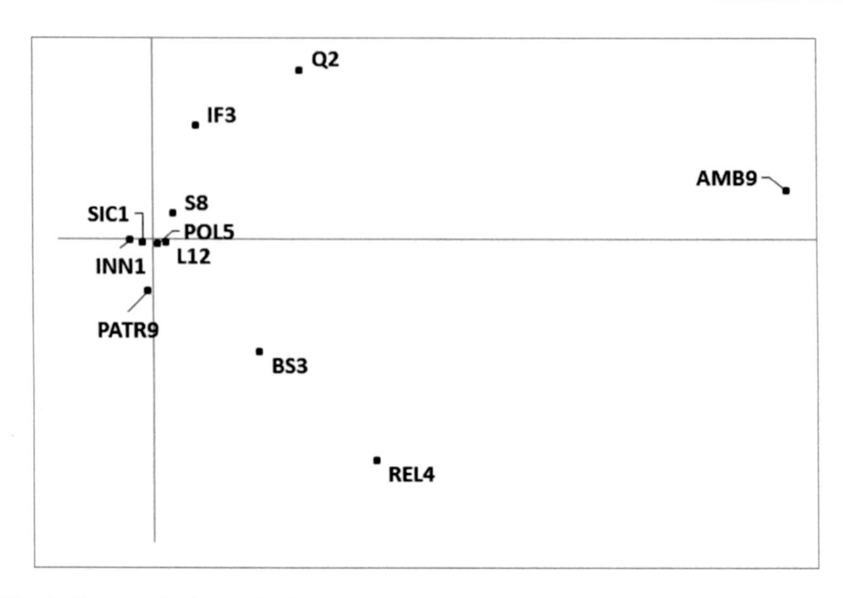

Fig. 2 Unconstrained analysis of the matrix Φ

disposable income. Table 4 reports the correlations between the variables in $\mathbf{Y}$ and the standardized predictors, as columns of the matrix Φ (see the diagonal of Table 4). The correlations in bold of the diagonal elements are the correlation between a single criterion variable $\mathbf{y}_q$ and its predicted values $\boldsymbol{\phi}_q$. When this type of correlation reports low values, this may indicate the major influence of the model covariates on the estimation of RCM random effects, and at the same time, highlights the scarce bond that exists between some of the components of the multivariate dependent variables and their model random effects. This is particularly clear for variables L12 (Share of employed persons who feel satisfied with their work) and POL5 (Trust in other institutions like the police and the fire brigade). These aspects are virtually "not explained" by the component of the variability of the random effects, due to the subspace by $\mathbf{X}^{\otimes}$ (the disposable income). To represent graphically the set of correlations, Fig. 3 reports the plot of the original variables in $\mathbf{Y}$ (dashed lines) and of predictor Φ (solid lines) onto the subspace of the first two principal components of $\mathbf{Y}$. The acronym STDP stands for "Standardized Predictor".

Finally, the analysis reflects correctly the state of the art of the dependence of the BES indicators with respect to the disposable income in Northern, Central, and Southern Italy (Fig. 1).

Table 4 Correlations between the variables (**Y**) and the standardized predictors (Φ), on the diagonal the correlation between the variable $i - th$ and its standardized predictor

Variable	AMB9	BS3	IF3	INN1	L12	PATR9	POL5	Q2	REL4	S8	SIC1
AMB9 STDP	**0.6979**	0.2486	0.1351	−0.5511	0.4772	−0.0307	0.0580	0.1762	0.3737	0.2070	−0.3504
BS3 STDP	0.0623	**0.6234**	−0.1868	−0.3057	0.0091	0.1230	−0.1269	−0.1515	0.0589	−0.2402	0.1085
IF3 STDP	−0.0889	−0.1623	**0.4742**	−0.1523	−0.3096	−0.1262	−0.4178	−0.1814	−0.2694	−0.0970	0.0302
INN1 STDP	−0.6042	−0.2083	−0.1193	**0.6110**	−0.5471	0.3078	−0.2496	−0.3410	−0.4105	−0.2506	0.3653
L12 STDP	0.0257	0.0178	−0.3483	−0.7358	**0.0239**	−0.2088	−0.3891	−0.4240	−0.1321	−0.4390	0.1346
PATR9 STDP	−0.0469	0.1153	−0.0760	0.2411	−0.1456	**0.9720**	−0.0554	−0.1616	0.0193	0.0540	−0.0476
POL5 STDP	−0.2114	−0.0336	−0.4769	−0.4673	−0.2300	−0.1030	**−0.0946**	−0.4705	−0.1797	−0.4763	0.2322
Q2 STDP	0.2422	0.0954	0.1433	−0.1558	0.1512	−0.2333	0.0138	**0.5288**	0.1017	0.1129	−0.1388
REL4 STDP	0.3461	0.2789	−0.0952	−0.3724	0.3633	0.0734	0.1157	−0.0310	**0.4956**	−0.0630	−0.1721
S8 STDP	0.1519	−0.1527	0.1294	−0.2213	−0.0907	0.1958	−0.2411	−0.1380	−0.1654	**0.1417**	−0.2627
SIC1 STDP	−0.4082	−0.0317	−0.2335	0.1182	−0.3123	−0.1368	−0.1748	−0.2323	−0.3065	−0.4109	**0.7470**

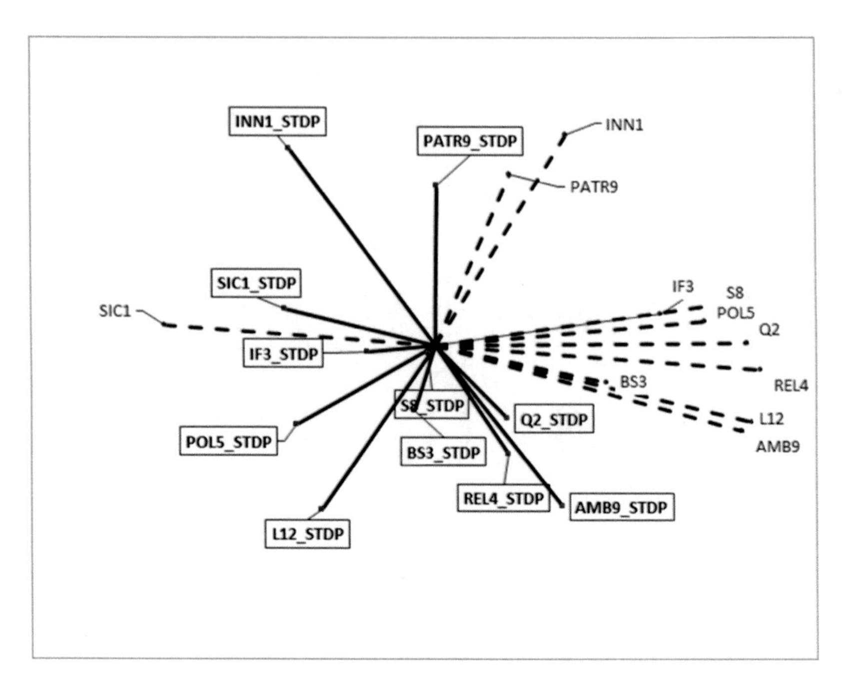

Fig. 3 Original variables (**Y**) and predictors (Φ) onto the subspace of the first two principal components of **Y**

4 Conclusions

Random-coefficients models represent an application of the general linear mixed models. In our opinion, one of the most important features of the RCM is that they allow adapting coefficients of the linear regression to independent clustered data. This reveals the nature of clusters, in what is common and what different from the other clusters belonging to the same set of data. When the response is multivariate, the common set of random coefficients of the regression model vary among clusters. One of the research issues may be related to the analysis of the contribution of the subspace of the model covariates to the variability of subjects. As reported in this study, by the application of the redundancy analysis to the standardized multivariate predictor, the independent variables of the model may reveal different types of influence, on subjects and on their own between-cluster variability.

5 Appendix

Proof Theorem 1—Conditional distribution of the multivariate random effects. We write the conditional density of $\mathbf{a}$ given $\mathbf{y}^*$ letting first $E(\mathbf{a}) = E(\mathbf{y}^*) = 0$, as

$$f(\mathbf{a}|\mathbf{y}^*) = \frac{f(\mathbf{a}, \mathbf{y}^*)}{f(\mathbf{y}^*)} = (2\pi)^{-\frac{rq}{2}} (|\Omega|/|\Omega_{\mathbf{y}^*,\mathbf{y}^*}|)^{-\frac{1}{2}}$$
$$\times \exp\left\{-\frac{1}{2}\boldsymbol{\alpha}'\Omega^{-1}\boldsymbol{\alpha} - \mathbf{y}^{*'}\Omega^{-1}_{\mathbf{y}^*,\mathbf{y}^*}\mathbf{y}^*\right\},$$

i.e., the ratio between the joint density of α and the marginal density of $\mathbf{y}^*$. Further, the determinant of Ω is $|\Omega| = |\Omega_{\mathbf{y}^*,\mathbf{y}^*}||\Omega_{a,a} - \Omega_{\mathbf{a},\mathbf{y}^*}\Omega^{-1}_{\mathbf{y}^*,\mathbf{y}^*}\Omega_{\mathbf{y}^*,\mathbf{a}}|$, with all the other determinants as positive numbers. The inverse of $\text{cov}(\boldsymbol{\alpha})$ is

$$\Omega^{-1} = \begin{bmatrix} \boldsymbol{C}_1 & -\boldsymbol{C}_1\Omega_{a,y^*}\boldsymbol{\Omega}^{-1}_{y^*,y^*} \\ -\boldsymbol{\Omega}^{-1}_{y^*,y^*}\boldsymbol{\Omega}_{y^*,a}\boldsymbol{C}_1 & \boldsymbol{C}_2 \end{bmatrix},$$
$$\boldsymbol{C}_1 = (\boldsymbol{\Omega}_{a,a} - \boldsymbol{\Omega}_{a,y^*}\boldsymbol{\Omega}^{-1}_{y^*,y^*}\boldsymbol{\Omega}_{y^*,a})^{-1},\ \ \boldsymbol{C}_2 = (\boldsymbol{\Omega}_{y^*,y^*} - \boldsymbol{\Omega}_{y^*,a}\boldsymbol{\Omega}^{-1}_{a,a}\boldsymbol{\Omega}_{a,y^*})^{-1}.$$

Then

$$f(\boldsymbol{a}|\boldsymbol{y}^*) = (2\pi)^{-\frac{rq}{2}}|\boldsymbol{C}_1|^{-\frac{1}{2}}\times$$

$$\exp\left\{-\frac{1}{2}[a - \boldsymbol{\Omega}_{a,y^*}\boldsymbol{\Omega}^{-1}_{y^*,y^*}y^*)'\boldsymbol{C}_1(a - \boldsymbol{\Omega}_{a,y^*}\boldsymbol{\Omega}^{-1}_{y^*,y^*}y^*) + y^{*\prime}(\boldsymbol{C}_2 - \boldsymbol{\Omega}^{-1}_{y^*,y^*})y^*]\right\},$$

where $\exp\left\{-\frac{1}{2}y^{*\prime}(\boldsymbol{C}_2 - \boldsymbol{\Omega}^{-1}_{y^*,y^*})y^*\right\}$ does not depend on $\boldsymbol{a}$. Substituting $\boldsymbol{a}$ with $\boldsymbol{a} - E(\boldsymbol{a})$, $\boldsymbol{y}^*$ with $\boldsymbol{y}^* - E(\boldsymbol{y}^*)$, and eliminating the last exponential after integration over $\boldsymbol{a}$, we have the formula of conditional distribution $\boldsymbol{a}|\boldsymbol{y}^*$. □

Proof Proposition 1—From the definition of the population model $\mathbf{y}_i = \mathbf{B}'\mathbf{x}_i + \mathbf{A}_i\mathbf{z}_i$, we say Ψ as the covariance matrix of the random vectors $(\widehat{\boldsymbol{\beta}} - \boldsymbol{\beta})$ and $\mathbf{Z}(\widetilde{\mathbf{a}} - \mathbf{a})$ of the *RCM* $\mathbf{Y} = \mathbf{X}^K\mathbf{B} + \mathbf{Z}^K\mathbf{A} + \mathbf{E}$, with $\mathbf{y}^* = \text{vec}(\mathbf{Y})$, $\mathbf{X} = (\mathbf{I} \otimes \mathbf{X}^{\otimes})$, and $\mathbf{Z} = (\mathbf{I} \otimes \mathbf{Z}^{\otimes}) = (\mathbf{I} \otimes \mathbf{I} \otimes \mathbf{X}_d)$. This matrix is as follows:

$$\Psi = \begin{bmatrix} \Psi_{11} & \Psi_{12} \\ \Psi'_{12} & \Psi_{22} \end{bmatrix} = \begin{bmatrix} \text{cov}(\widehat{\boldsymbol{\beta}} - \boldsymbol{\beta}) & \text{cov}[\widehat{\boldsymbol{\beta}} - \boldsymbol{\beta}, \mathbf{Z}(\widetilde{\boldsymbol{a}} - \boldsymbol{a})] \\ \left\{\text{cov}[\widehat{\boldsymbol{\beta}} - \boldsymbol{\beta}, \mathbf{Z}(\widetilde{\boldsymbol{a}} - \boldsymbol{a})]\right\}' & \text{cov}[\mathbf{Z}(\widetilde{\boldsymbol{a}} - \boldsymbol{a})] \end{bmatrix}.$$

Consider now the Best Linear Unbiased Estimator of $\boldsymbol{\beta}$

$$\widehat{\boldsymbol{\beta}} = (\mathbf{X}'\text{cov}(\mathbf{y}^*)^{-1}\mathbf{X})^{-1}\mathbf{X}'\text{cov}(\mathbf{y}^*)^{-1}\mathbf{y}$$

and the estimator $E(\mathbf{a}|\mathbf{y}^*) = \widetilde{\mathbf{a}} = \mathbf{D}\mathbf{Z}'\text{cov}(\mathbf{y}^*)^{-1}(\mathbf{y} - \mathbf{X}\widehat{\boldsymbol{\beta}})$ of $\mathbf{a}$, with $\text{cov}(\mathbf{y}^*) = \mathbf{Z}\mathbf{D}\mathbf{Z}' + \mathbf{R}$ and $\text{cov}(\widehat{\boldsymbol{\beta}} - \boldsymbol{\beta}) = \text{cov}(\widehat{\boldsymbol{\beta}})$.

Let us write $\mathbf{P}_X = \mathbf{X}(\mathbf{X}'\text{cov}(\mathbf{y}^*)^{-1}\mathbf{X})^{-1}\mathbf{X}'\text{cov}(\mathbf{y}^*)^{-1}$ as the projection matrix in the $q \times p$ -dimensional subspace given by the column vectors of $\mathbf{X}$ in the metric $\text{cov}(\mathbf{y}^*)^{-1}$, and $\mathbf{P} = \text{cov}(\mathbf{y}^*)^{-1}(\mathbf{I} - \mathbf{P}_X)$ as the projection matrix onto the orthogonal complement of the column space of $\mathbf{X}$. Assuming $\boldsymbol{\theta} = \boldsymbol{\theta}^*$, $\text{cov}((\widehat{\boldsymbol{\beta}} - \boldsymbol{\beta}) + \mathbf{Z}(\widetilde{\mathbf{a}} - \mathbf{a})|\boldsymbol{\theta}^*) = \text{cov}[\widehat{\boldsymbol{\beta}} + \mathbf{Z}(\widetilde{\mathbf{a}} - \mathbf{a})]$, and given $\mathbf{Q} = (\mathbf{X}'\text{cov}(\mathbf{y}^*)^{-1}\mathbf{X})^{-1}\mathbf{X}'\text{cov}(\mathbf{y}^*)^{-1}$, we have that:

$$\begin{aligned}\Psi_{22} &= \text{cov}[\mathbf{Z}(\widetilde{\mathbf{a}} - \mathbf{a})] = \mathbf{Z}[\text{cov}(\widetilde{\mathbf{a}}) + \text{cov}(\mathbf{a}) - \text{cov}(\widetilde{\mathbf{a}}, \mathbf{a}) - \text{cov}(\mathbf{a}, \widetilde{\mathbf{a}})]\mathbf{Z}' \\ &= \mathbf{Z}[\mathbf{DZ}'\mathbf{PZD} + \text{cov}(\mathbf{a}) - \text{cov}(\widetilde{\mathbf{a}}) - \text{cov}(\widetilde{\mathbf{a}})]\mathbf{Z}' = \mathbf{ZDZ}' - \mathbf{ZDZ}'\mathbf{PZDZ}',\end{aligned}$$

where $\text{cov}(\widetilde{\mathbf{a}}) = \text{cov}(\mathbf{a}|\mathbf{y}^*) = \text{cov}[\mathbf{DZ}'\text{cov}(\mathbf{y}^*)^{-1}(\mathbf{y} - \mathbf{X}\widehat{\boldsymbol{\beta}})] = \mathbf{DZ}'\mathbf{PZD}$, and $\text{cov}(\widetilde{\mathbf{a}}, \mathbf{a}) = \text{cov}[\mathbf{DZ}'\text{cov}(\mathbf{y}^*)^{-1}(\mathbf{y} - \mathbf{X})\widehat{\boldsymbol{\beta}}), \mathbf{a}] = \text{cov}(\widetilde{\mathbf{a}})$.

Since $\mathbf{Q}\text{var}(\mathbf{y}^*)\mathbf{P} = \mathbf{Q}\text{cov}(\mathbf{y}^*, \mathbf{y}^*)\mathbf{P} = \mathbf{0}$, thus we get

$$\begin{aligned}\Psi_{12} &= \text{cov}[\widehat{\boldsymbol{\beta}}, \mathbf{Z}(\widetilde{\mathbf{a}} - \mathbf{a})] = \mathbf{Q}[\text{cov}(\mathbf{y}^*, \widetilde{\mathbf{a}}) - \text{cov}(\mathbf{y}^*, \mathbf{a})]\mathbf{Z}' \\ &= \mathbf{Q}[\text{cov}(\mathbf{y}^*, \mathbf{DZ}'\mathbf{Py}^*) - \mathbf{ZD}]\mathbf{Z}' = \mathbf{Q}[\text{var}(\mathbf{y}^*)\mathbf{PZD} - \mathbf{ZD}]\mathbf{Z}' \\ &= -\mathbf{QZDZ}' = -(\mathbf{X}'\text{cov}(\mathbf{y}^*)^{-1}\mathbf{X})^{-1}\mathbf{X}'\text{cov}(\mathbf{y}^*)^{-1}(\mathbf{I} \otimes \mathbf{X}_d)\mathbf{D}(\mathbf{I} \otimes \mathbf{X}_d').\end{aligned}$$

Finally, given $\mathbf{e}^* = \text{vec}(\mathbf{E})$, we get:

$$\begin{aligned}\Psi_{11} &= \text{cov}(\widehat{\boldsymbol{\beta}} - \boldsymbol{\beta}) = \text{cov}(\widehat{\boldsymbol{\beta}}) = E[\mathbf{Qy}^*(\mathbf{y}^*)'\mathbf{Q}'] \\ &= \mathbf{Q}E[\mathbf{Zaa}'\mathbf{Z}' + \mathbf{e}^*(\mathbf{e}^*)']\mathbf{Q}' = [\mathbf{X}'(\mathbf{ZDZ}' + \mathbf{R})^{-1}\mathbf{X}]^{-1}.\end{aligned}$$

This ends the proof of the Proposition.

Proof Theorem 2—Rank of $\boldsymbol{\Phi}$.
Remembering that, in the usual applications of the linear model $\boldsymbol{\Phi}$, the condition $n \gg p \gg h$ is the more realistic, first observe for (a): $\boldsymbol{r}(\boldsymbol{XB}) \leq \min[\boldsymbol{r}(\boldsymbol{X}), \boldsymbol{r}(\boldsymbol{B})]$. If matrices $\boldsymbol{X}$ and $\boldsymbol{B}$ are of full rank, then $\boldsymbol{r}(\boldsymbol{X}) = \boldsymbol{r}(\boldsymbol{B}) = h$. That is, the full column rank of $\boldsymbol{X}$ equals the full row rank of $\boldsymbol{B}$. For the verification of (b), if $\boldsymbol{\alpha}$ is a matrix of m zero-mean independent p-variate row random vectors ($E(\boldsymbol{\alpha}_i') = \mathbf{0}_{p\times 1}$), with $\boldsymbol{Z}$ a $n \times m$ matrix, then by model definition matrix $\boldsymbol{\alpha}'\boldsymbol{Z}'$ form the homogeneous system $\boldsymbol{\alpha}'\boldsymbol{Z}'\mathbf{1}_n = \mathbf{0}_{p\times 1}$, with the nontrivial solution $\mathbf{1}_n$. In fact, with $p > m$ and $m = \min[\boldsymbol{r}(\boldsymbol{Z}), \boldsymbol{r}(\boldsymbol{\alpha})]$, we have that $(\boldsymbol{\alpha}'\boldsymbol{Z}')^{-}\boldsymbol{\alpha}'\boldsymbol{Z}' \neq \boldsymbol{I}_n$, and, consequently, $\boldsymbol{r}(\boldsymbol{Z\alpha}) < m$. Then, if the set of all solution of $\boldsymbol{\alpha}'\boldsymbol{Z}'\mathbf{1}_n = 0$ represents the null space of $\boldsymbol{\alpha}'\boldsymbol{Z}'$, with $\dim[\mathcal{N}(\boldsymbol{\alpha}'\boldsymbol{Z}')] = 1$, it follows that $\boldsymbol{r}(\boldsymbol{Z\alpha}) = m - \dim[\mathcal{N}(\boldsymbol{\alpha}'\boldsymbol{Z}')] = m - 1$.

For the statements in (c) and (d), first observe that $\boldsymbol{\eta}$ is a $n \times p$ matrix of full column rank, with independent row and column random vectors, where $p \ll n$, $\boldsymbol{r}(\boldsymbol{\eta}) = p$. Further $\text{cov}(\boldsymbol{\alpha}_i, \boldsymbol{\eta}_i) = 0$ and $\text{cov}[\text{vec}(\boldsymbol{\Phi}), \text{vec}(\boldsymbol{\alpha})] = (\boldsymbol{I}_p \otimes \mathbf{Z})(\boldsymbol{\Sigma}_{\alpha} \otimes \boldsymbol{I}_m) = \boldsymbol{\Sigma}_{\alpha} \otimes \mathbf{Z}$, independent of $\boldsymbol{X}$. Thus, for column spaces $\mathcal{C}(\boldsymbol{XB}), \mathcal{C}(\boldsymbol{Z\alpha})$, $\dim[\mathcal{C}(\boldsymbol{XB}) \cap \mathcal{C}(\boldsymbol{Z\alpha})] = 0$ and, in conclusion,

$$\boldsymbol{r}(\boldsymbol{XB}) + \boldsymbol{r}(\boldsymbol{Z\alpha}) = \min[\boldsymbol{r}(\boldsymbol{X}), \boldsymbol{r}(\boldsymbol{B})] + \min[\boldsymbol{r}(\boldsymbol{Z}), \boldsymbol{r}(\boldsymbol{\alpha})] = h + m.$$

To prove (d), note that

(i) from the column space $\mathcal{C}(\boldsymbol{\eta}) = p$, and $\dim[\mathcal{C}(\boldsymbol{XB}) \cap \mathcal{C}(\boldsymbol{Z\alpha}) \cap \mathcal{C}(\boldsymbol{\eta})] = 0$, it follows that $\boldsymbol{r}(\boldsymbol{\Phi}) = \boldsymbol{r}(\boldsymbol{XB}) + \boldsymbol{r}(\boldsymbol{Z\alpha}) + \boldsymbol{r}(\boldsymbol{\eta})$.
(ii) $\boldsymbol{r}(\boldsymbol{\eta}) \leq \min(n, p)$. Thus $\boldsymbol{r}(\boldsymbol{\eta}) = \boldsymbol{r}(\boldsymbol{\Phi}) = p$.

Validation of (e) follows immediately, after noting that

$$\boldsymbol{r}(\boldsymbol{XB}) + \boldsymbol{r}(\boldsymbol{Z\alpha}) \leq \min(h + m, p).$$

□

References

Demidenko, E. (2004). *Mixed models: Theory and applications*. Wiley.
Härdle, W. K., & Simar, L. (2015). Principal components analysis. In *Applied multivariate statistical analysis* (pp. 319–358). Berlin: Springer.
Hsiao, C., & Pesaran, M. H. (2008). Random coefficient models. In L. Mátyás, & P. Sevestre (Eds.), The econometrics of panel data. Advanced studies in theoretical and applied econometrics (Vol. 46). Berlin: Springer.
Ledoit, O., & Wolf, M. (2004). Honey, I shrunk the sample covariance matrix. *Journal of Portfolio Management, 31*(1), 110–119.
Longford, N. T. (2011). Random coefficient models. In M. Lovric (Ed.), *International encyclopedia of statistical science*. Berlin: Springer.
Marcis, L., & Salvatore, R. (2020). Joint redundancy analysis by a multivariate linear predictor. In *Conference of the Italian Statistical Society (SIS2020)*.
Takane, Y., & Hwang, H. (2007). Regularized linear and kernel redundancy analysis. *Computational Statistics and Data Analysis, 52*(1), 394–405.
Timm, N. H. (2002). *Applied multivariate analysis*. Springer.
Ploymukda, Arnon, Chansangiam, Pattrawut, & Wicharn, Lewkeeratiyutkul. (2018). Analytic properties of Tracy-Singh products for operator matrices. *Journal of Computational Analysis and Applications, 24*, 665–674.
Van Den Wollenberg, A. L. (1977). Redundancy analysis an alternative for canonical correlation analysis. *Psychometrika, 42*(2), 207–219.

Parsimonious Mixtures of Matrix-Variate Shifted Exponential Normal Distributions

Salvatore D. Tomarchio, Luca Bagnato, and Antonio Punzo

Abstract Finite mixtures of matrix-variate distributions constitute a powerful model-based clustering device. One serious issue of these models is the potentially high number of parameters to be estimated. Thus, in this work we introduce a family of 196 parsimonious mixture models based on the matrix-variate shifted exponential normal distribution, an elliptical heavy-tailed generalization of the matrix-variate normal distribution. Parsimony is introduced in a twofold manner: (i) by using the eigendecomposition of the components scale matrices and (ii) by allowing the components tailedness parameter to be tied across the groups. A further characteristic of the proposed models relies on the more flexible tail behavior with respect to existing parsimonious matrix-variate normal mixtures, thus allowing for a better modeling of datasets having atypical observations. Parameter estimation is obtained by using an ECM algorithm. The proposed models are then fitted to a real dataset along with parsimonious matrix-variate normal mixtures for comparison purposes.

Keywords Mixture models · Matrix-variate · Clustering · Parsimony

1 Introduction

Finite mixture models are the basis of many modern clustering algorithms. Within this context, finite mixtures for matrix-variate data have received particular attention

S. D. Tomarchio (✉) · A. Punzo
Dipartimento di Economia e Impresa, Università degli Studi di Catania, Catania, Italia
e-mail: daniele.tomarchio@unict.it

A. Punzo
e-mail: antonio.punzo@unict.it

L. Bagnato
Dipartimento di Scienze Economiche e Sociali, Università Cattolica del Sacro Cuore, Piacenza, Italia
e-mail: luca.bagnato@unicatt.it

L. Grilli et al. (eds.), *Statistical Models and Methods for Data Science*, Studies in Classification, Data Analysis, and Knowledge Organization,
https://doi.org/10.1007/978-3-031-30164-3_14

in the recent statistical literature (Melnykov et al. 2018; Gallaugher and McNicholas 2018; Melnykov et al. 2019; Tomarchio et al. 2020, 2022, 2021; Sarkar et al. 2020). As discussed by Viroli (2011), many complex data structures can be arranged in a matrix-variate (three-way) form factor, and depending on the entity indexed in each of the three layers, different data examples may be obtained.

A common issue of matrix-variate mixture models is the potentially high number of parameters to be estimated. This is mainly related to the dimensionality of the two scale matrices of each mixture component. Indeed, for a sample of $p \times r$-dimensional matrices, a total of $p(p+1)/2 + r(r+1)/2 - 1$ unique parameters related to the scale matrices must be estimated. One solution to this problem consists in the use of the well-known eigendecomposition of the component covariance matrices (Celeux and Govaert 1995), as recently done for matrix-variate normal mixtures (MVN-Ms) by Sarkar et al. (2020). By following this approach, in this work we introduce a family of parsimonious mixture models based on the matrix-variate shifted exponential normal (MVSEN) distribution, an elliptical heavy-tailed generalization of the matrix-variate normal (MVN) distribution, recently introduced in the literature in Tomarchio et al. (2020). To further increase the parsimony of our models, we also allow the option of constraining the tailedness parameter of each mixture component to be equal. By combining these two sources of parsimony, we obtain a total of 196 parsimonious matrix-variate shifted exponential normal mixtures (MVSEN-Ms). An advantage of the proposed models with respect to MVN-Ms is the greater flexibility in terms of tail behavior of the mixture components. Indeed, for many real datasets, the tails of the MVN distribution are lighter than required, with direct consequences to the parameter estimation process and to the recovery of the underlying group structure.

The paper is organized as follows. In Sect. 2, we firstly present our family of models and then we discuss an expectation-conditional maximization (ECM) algorithm (Meng and Rubin 1993) for maximum likelihood parameter estimation. In Sect. 3, we apply our family of models to a real dataset concerning R&D-performing US manufacturing companies observed for 8 years. Finally, some conclusions and ideas for future developments are given in Sect. 4.

2 Methodology

2.1 *Parsimonious Mixtures of Matrix-Variate Shifted Exponential Normal Distributions*

A $p \times r$-dimensional random matrix $\mathcal{X}$ arises from a finite mixture model if its probability distribution function (pdf) is

$$p(\mathbf{X}; \Omega) = \sum_{g=1}^{G} \pi_g f(\mathbf{X}; \Theta_g), \tag{1}$$

where π_g is the mixing proportion for the gth component, with $\pi_g > 0$ and $\sum_{g=1}^{G} \pi_g = 1$, $f(\mathbf{X}; \Theta_g)$ is the pdf of the gth component having parameter Θ_g, and Ω contains all of the parameters of the mixture.

Herein, we consider the MVSEN distribution as functional form of the mixture components in (1). We recall that the pdf of the MVSEN distribution is

$$f(\mathbf{X}; \Theta) = \frac{\theta \exp(\theta)}{(2\pi)^{\frac{pr}{2}} |\Sigma|^{\frac{r}{2}} |\Psi|^{\frac{p}{2}}} \varphi_{\frac{pr}{2}} \left(\frac{\delta(\mathbf{X}; \mathbf{M}, \Sigma, \Psi)}{2} + \theta \right), \tag{2}$$

where $\delta(\mathbf{X}; \mathbf{M}, \Sigma, \Psi) = \text{tr}\left[\Sigma^{-1}(\mathbf{X} - \mathbf{M})\Psi^{-1}(\mathbf{X} - \mathbf{M})'\right]$, $\mathbf{M}$ is the $p \times r$ mean matrix, Σ and Ψ are the $p \times p$ and $r \times r$ row and column scale matrices, respectively, $\theta > 0$ is the tailedness parameter and $\varphi_m(z)$ is the Misra function (Misra 1940), generalized form of the exponential integral function. We also remind that the MVN distribution is a limiting case of the MVSEN distribution when $\theta \to \infty$; for further details see Tomarchio et al. (2020) and appendix B of Punzo and Bagnato (2020) that can be easily generalized to show this relationship.

For parameter estimation, it is convenient to write the pdf in (2) by using its hierarchical representation obtained via the matrix-variate normal scale mixture model (Tomarchio et al. 2020)

1. $W \sim \mathcal{SE}(\theta)$,
2. $\mathbf{X}|W = w \sim \mathcal{N}(\mathbf{M}, \Sigma/w, \Psi)$,

where W is a positive mixing random variable, $\mathcal{SE}(\cdot)$ denotes a shifted exponential distribution on $(1, +\infty)$, and $\mathcal{N}(\cdot)$ is the MVN distribution.

Finite mixtures of MVSEN distributions have been introduced in an unconstrained setting in Tomarchio et al. (2020). However, as discussed in Sect. 1, they might suffer of overparameterization issues. To address this concern, we implement the following two sources of parsimony:

1. the eigendecomposition of the components scale matrices Σ and Ψ. By recalling Celeux and Govaert (1995), we have that a generic $q \times q$ component scale matrix Φ_g can be decomposed as

$$\Phi_g = \lambda_g \Gamma_g \Delta_g \Gamma_g', \tag{3}$$

where $\lambda_g = |\Phi_g|^{1/q}$, Γ_g is a $q \times q$ orthogonal matrix whose columns are the normalized eigenvectors of Φ_g, and Δ_g is the scaled ($|\Delta_g| = 1$) diagonal matrix of the eigenvalues of Φ_g. These elements correspond respectively to the volume, orientation, and shape of the gth cluster. By constraining the three components in (3), the following 14 parsimonious structures are obtained: EII, VII, EEI, VEI, EVI, VVI, EEE, VEE, EVE, VVE, EEV, VEV, EVV, VVV, where "E" stands for equal, "V" means varying, and "I" denotes the identity matrix.
Notice that we do not obtain 14 parsimonious structures for each scale matrix. Indeed, to address an identifiability issue related to the scale matrices, we impose the restriction $|\Psi_g| = 1$. This makes the λ_g parameter unnecessary in the decomposition of Ψ_g, reducing for this matrix the parsimonious structures from 14 to 7:

II, EI, VI, EE, VE, EV, VV. Therefore, we globally obtain a total of $14 \times 7 = 98$ parsimonious structures for the component scale matrices.
2. The tailedness parameter θ_g of the mixture components is allowed to be equal among them. The nomenclature used is the same as that adopted for the scale matrices, i.e., "E" refers to tied tailedness parameters across groups and "V" is used for the unconstrained case.

By combining these two sources of parsimony, we obtain a family of $98 \times 2 = 196$ parsimonious MVSEN mixture models.

2.2 Maximum Likelihood Estimation

To estimate the parameters of our family of mixture models we adopt the ECM algorithm, an extension of the expectation-maximization (EM) algorithm (Dempster et al. 1977), which is a natural approach for maximum likelihood estimation when data are incomplete. By using the hierarchical representation of Sect. 2.1 we have two sources of incompleteness:

1. the first is due to the unknown cluster membership of each observation;
2. the second is due to the mixing random variable W.

Therefore, the complete data are $\mathbf{S}_c = \{\mathbf{X}_i, \mathbf{z}_i, w_i\}_{i=1}^N$, where $\mathbf{z}_i = (z_{i1}, \dots, z_{iG})'$, such that $z_{ig} = 1$ if observation i belongs to group g and $z_{ig} = 0$ otherwise, and w_i is the realization of W. Because of this conditional structure, the complete data log likelihood can be written as

$$\ell_c(\Omega; \mathbf{S}_c) = \ell_{1c}(\boldsymbol{\pi}; \mathbf{S}_c) + \ell_{2c}(\Xi; \mathbf{S}_c) + \ell_{3c}(\boldsymbol{\vartheta}; \mathbf{S}_c), \tag{4}$$

where

$$\ell_{1c}(\boldsymbol{\pi}; \mathbf{S}_c) = \sum_{i=1}^{N} \sum_{g=1}^{G} z_{ig} \ln(\pi_g), \tag{5}$$

with $\boldsymbol{\pi} = \{\pi_g\}_{g=1}^G$,

$$\begin{aligned}\ell_{2c}(\Xi; \mathbf{S}_c) = \sum_{i=1}^{N} \sum_{g=1}^{G} z_{ig} \Bigg[& -\frac{pr}{2}\ln(2\pi) + \frac{pr}{2}\ln(w_{ig}) - \frac{r}{2}\ln|\Sigma_g| - \frac{p}{2}\ln|\Psi_g| \\ & - \frac{w_{ig}\delta_g(\mathbf{X}; \mathbf{M}_g, \Sigma_g, \Psi_g)}{2}\Bigg],\end{aligned} \tag{6}$$

where $\Xi = \{\mathbf{M}_g, \Sigma_g, \Psi_g\}_{g=1}^G$ and

$$\ell_{3c}\left(\boldsymbol{\vartheta}; \mathbf{S}_c\right) = \sum_{i=1}^{N}\sum_{g=1}^{G} z_{ig}\left[\ln\left(\theta_g\right) - \theta_g\left(w_{ig} - 1\right)\right], \tag{7}$$

with $\boldsymbol{\vartheta} = \left\{\theta_g\right\}_{g=1}^{G}$. Then, by marking with one dot the parameters updates at the previous iteration and with two dots those related to the current iteration, the ECM algorithm proceeds as follows.

E-Step At the E-step, it is necessary to compute

$$\ddot{z}_{ig} := E\left(Z_{ig}; \mathbf{X}_i, \dot{\Omega}\right) = \frac{\dot{\pi}_g f\left(\mathbf{X}_i; \dot{\Theta}_g\right)}{\sum_{h=1}^{G} \dot{\pi}_h f\left(\mathbf{X}_i; \dot{\Theta}_h\right)},$$

$$\ddot{w}_{ig} := E\left(W_{ig}; \mathbf{X}_i, z_{ig} = 1, \dot{\Omega}\right) = \frac{\varphi_{\frac{pr}{2}+1}\left(\frac{\dot{\delta}_g(\mathbf{X}; \dot{\mathbf{M}}_g, \dot{\Sigma}_g, \dot{\Psi}_g)}{2} + \dot{\theta}_g\right)}{\varphi_{\frac{pr}{2}}\left(\frac{\dot{\delta}_g(\mathbf{X}; \dot{\mathbf{M}}_g, \dot{\Sigma}_g, \dot{\Psi}_g)}{2} + \dot{\theta}_g\right)}, \tag{8}$$

which are the posterior probability that the unlabeled observation $\mathbf{X}_i$ belongs to the gth mixture component and the expected value of a left-truncated gamma distribution with parameters $(pr/2) + 1$ and $\dot{\delta}_g(\mathbf{X}; \dot{\mathbf{M}}_g, \dot{\Sigma}_g, \dot{\Psi}_g)/2 + \dot{\theta}_g$ over the interval $(1, +\infty)$, respectively.

Now, consider $\Omega_1 = \left\{\pi_g, \mathbf{M}_g, \Sigma_g, \theta_g\right\}_{g=1}^{G}$ and $\Omega_2 = \left\{\Psi_g\right\}_{g=1}^{G}$.

CM-Step 1 At the first CM-step, keeping fixed Ω_2 at $\dot{\Omega}_2$, we maximize the expectation of the complete data log likelihood with respect to Ω_1. This results in the following parameter updates

$$\ddot{\pi}_g = \frac{\sum_{i=1}^{N} \ddot{z}_{ig}}{N}, \quad \ddot{\mathbf{M}}_g = \frac{\sum_{i=1}^{N} \ddot{z}_{ig}\ddot{w}_{ig}\mathbf{X}_i}{\sum_{i=1}^{N} \ddot{z}_{ig}\ddot{w}_{ig}}. \tag{9}$$

The update for θ_g depends on the choice of constraining or not its value to be tied across groups. In detail, they are

$$\ddot{\theta} = \frac{\sum_{i=1}^{N}\sum_{g=1}^{G} \ddot{z}_{ig}}{\sum_{i=1}^{N}\sum_{g=1}^{G} \ddot{z}_{ig}\left(\ddot{w}_{ig} - 1\right)}, \quad \ddot{\theta}_g = \frac{\sum_{i=1}^{N} \ddot{z}_{ig}}{\sum_{i=1}^{N}\left(\ddot{w}_{ig} - 1\right)}, \tag{10}$$

for the "E" and "V" cases, respectively.

Similarly, the update for Σ_g depends on the parsimonious structure considered. Because of space constraints, an exhaustive description on how to obtain the updates of each parsimonious model is not possible. However, useful details are provided in Sarkar et al. (2020). The difference with respect to results of Sarkar et al. (2020) consists in the update of the row scatter matrix associated with the gth component, that in our work is defined as

$$\ddot{\mathbf{W}}_g^R = \sum_{i=1}^{N} \ddot{z}_{ig} \ddot{w}_{ig} \left(\mathbf{X}_i - \ddot{\mathbf{M}}_g\right) \dot{\Psi}_g^{-1} \left(\mathbf{X}_i - \ddot{\mathbf{M}}_g\right)'. \tag{11}$$

CM-Step 2 At the second CM-step, keeping fixed Ω_1 at $\ddot{\Omega}_1$, we maximize the expectation of the complete data log likelihood with respect to Ω_2. Also in this case, the updates related to Ψ_g depend on the parsimonious structure considered, and for the same reasons mentioned above they are not reported here. Useful details can be found in the supplementary appendix of Sarkar et al. (2020). The difference with respect to results therein reported consists in the update of the column scatter matrix associated with the gth component, that in our work is defined as

$$\ddot{\mathbf{W}}_g^C = \sum_{i=1}^{N} \ddot{z}_{ig} \ddot{w}_{ig} \left(\mathbf{X}_i - \ddot{\mathbf{M}}_g\right)' \ddot{\Sigma}_g^{-1} \left(\mathbf{X}_i - \ddot{\mathbf{M}}_g\right). \tag{12}$$

3 Real Data Example

In this section, we analyze the `RDPerfComp` dataset contained in the **pder** package (Croissant and Millo 2019) for the R statistical software. This dataset consists of $p = 3$ variables measured for $N = 509$ R&D-performing US manufacturing companies over $r = 8$ years (1982–1989). Specifically, the logarithm of capital stock and employment are measured at the end of the firm's accounting year, while the logarithm of sales is used as a proxy for output.

We fitted to this dataset parsimonious MVSEN-Ms and MVN-Ms for $G \in \{1, \ldots, 6\}$, and for each family of models the Bayesian information criterion (BIC) (Schwarz 1978) and the integrated completed likelihood (ICL) (Biernacki et al. 2000) are used to select the best fitting model. The results are displayed in Table 1.

From the analysis of the results, we see that the parsimonious structure, as well as the number of groups G, detected by the BIC and the ICL are the same within each family of models. Furthermore, these results disclose the benefit of considering parsimony and the non-necessity of making the model overparametrized by fitting unconstrained mixtures. Anyway, MVSEN-Ms provide a better fitting than MVN-Ms according to both information criteria.

Table 1 Parsimonious structure, number of groups G, and value of the information criteria for the best fitting models, according to BIC and ICL, for each family of models

Family	BIC			ICL		
	Pars.	G	Value	Pars.	G	Value
MVN-Ms	VVE-VE	5	−9415.07	VVE-VE	5	−9365.13
MVSEN-Ms	VEE-VE-E	4	−10174.00	VEE-VE-E	4	−10075.40

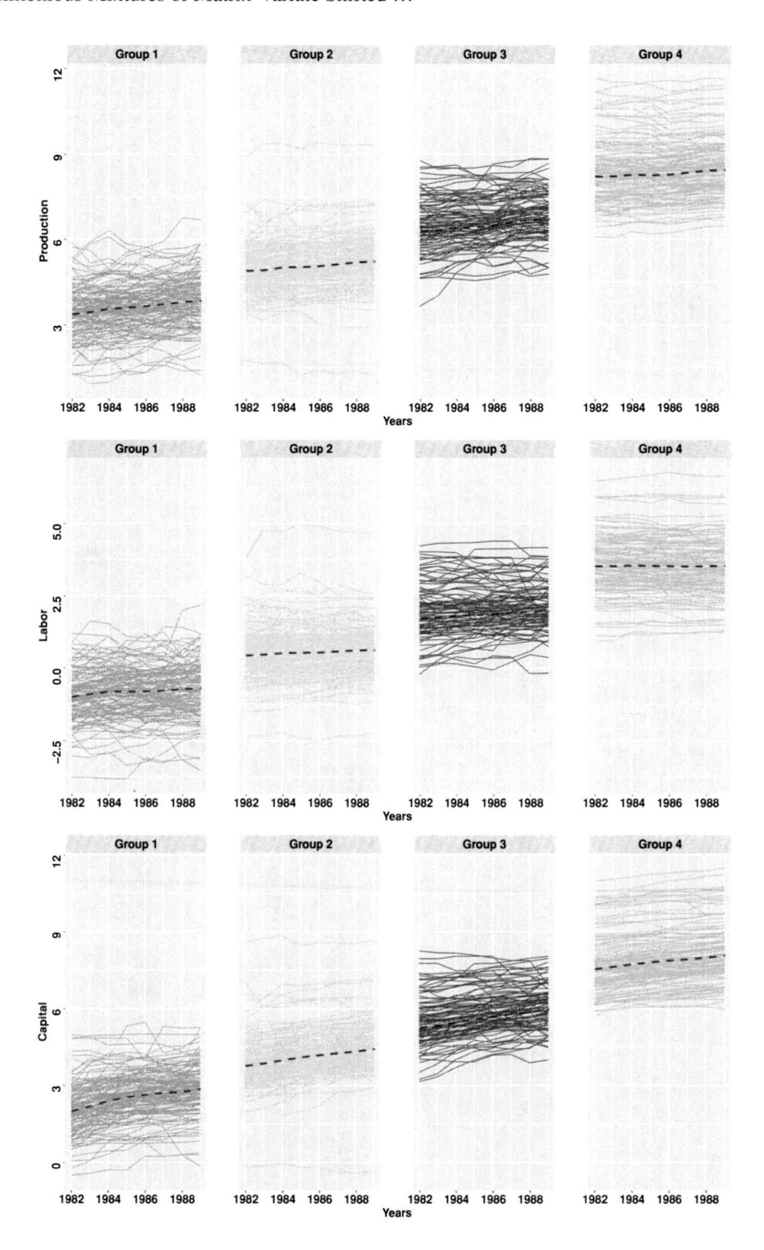

Fig. 1 Parallel coordinate plots of the data partition obtained via the VEE-VE-E MVSEN-Ms. The dashed lines are the estimated means

Notice also that MVN-Ms find an additional group in the data, and this might be an indication that the tails of the components MVN distributions are not heavy enough to adequately model this data. In this regard, useful indications can be obtained by

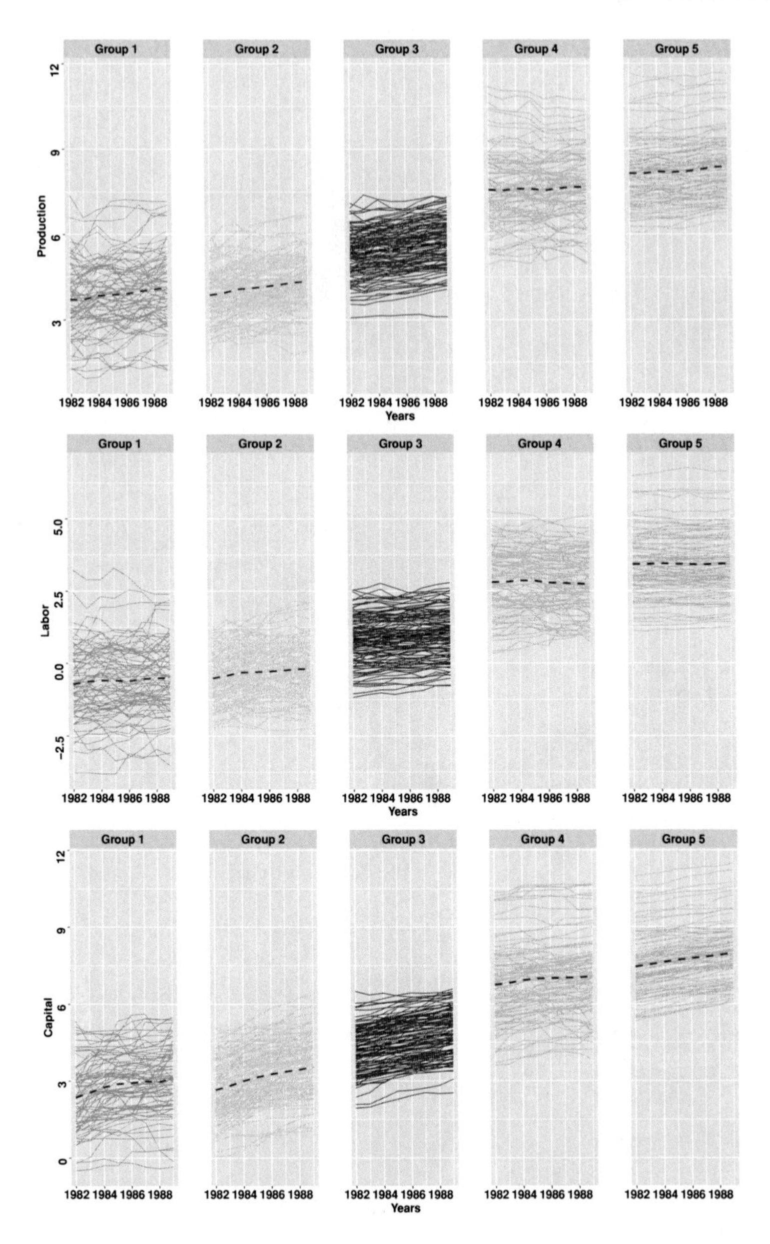

Fig. 2 Parallel coordinate plots of the data partition obtained via the VVE-VE MVN-Ms. The dashed lines are the estimated means

Table 2 Confusion matrix for the data partitions produced by the VEE-VE-E MVSEN-Ms and VVE-VE MVN-Ms

	MVN-Ms					
	G	1	2	3	4	5
MVSEN-Ms	1	69	58	2	0	0
	2	7	51	92	8	0
	3	4	4	31	51	8
	4	0	0	1	37	86

the analysis of the estimated data partitions. Specifically, Fig. 1 plots the parallel coordinate plots for the data partition detected by the VEE-VE-E MVSEN-Ms. The dashed lines represent the estimated means for that variable, over time, in that group. We sorted the estimated groups according to the values assumed by the estimated mean matrices. By looking at the estimated means, we can say that the detected groups are most of the times fairly separated, and as we move from the first group to the fourth group the manufacturing companies grow in terms of production, labor, and capital involved. Conversely, Fig. 2 illustrates the parallel coordinate plots for the data partition produced by the VVE-VE MVN-Ms. Here, the estimated groups are much more overlapped, as also confirmed by the values of the estimated mean matrices that most of the times are very close to each other. If we compute the confusion matrix associated with the two data partitions, reported in Table 2, we see that the observations contained in each of the groups detected by MVSEN-Ms are roughly splitted among two groups, thus resulting in a quite different data classification.

4 Conclusions

Matrix-variate mixture models are often characterized by a high number of parameters to be estimated. Additionally, many real datasets require more flexible models than matrix-variate mixtures. To jointly consider both aspects, in this paper a family of 196 parsimonious mixture models, based on the matrix-variate shifted exponential normal distribution, is introduced. We introduce parsimony in our models via the eigendecomposition of the components scale matrices and by constraining the tailedness parameter of the mixture components to be tied. An ECM algorithm is discussed for parameter estimation. Our family of models have been fitted to a real dataset along with parsimonious matrix-variate normal mixtures. The results demonstrate the best fitting behavior of our models. Additionally, we reported that matrix-variate normal mixtures detected an additional group in the data, which may be an indication that the tails of its components distributions are not heavy enough to adequately model this data. Indeed, from the assessment of the estimated groups, those provided by matrix-variate normal mixtures are much more overlapped than those obtained by our (and overall) best fitting model.

Acknowledgements Antonio Punzo and Salvatore D. Tomarchio have been partially supported by MIUR, grant number 2022XRHT8R—*The SMILE project: Statistical Modelling and Inference to Live the Environment*.
This research also contributes to the PNRR GRInS Project.

References

Biernacki, C., Celeux, G., & Govaert, G. (2000). Assessing a mixture model for clustering with the integrated completed likelihood. *IEEE Transactions on Pattern Analysis and Machine Intelligence, 22*(7), 719–725.

Celeux, G., & Govaert, G. (1995). Gaussian parsimonious clustering models. *Pattern Recognition, 28*(5), 781–793.

Croissant, Y., & Millo, G. (2019). pder: Panel data econometrics with R. R package version 1.0-1.

Dempster, A. P., Laird, N. M., & Rubin, D. B. (1977). Maximum likelihood from incomplete data via the EM algorithm. *Journal of the Royal Statistical Society: Series B, 39*(1), 1–22.

Gallaugher, M. P. B., & McNicholas, P. D. (2018). Finite mixtures of skewed matrix variate distributions. *Pattern Recognition, 80*, 83–93.

Melnykov, V., & Zhu, X. (2018). On model-based clustering of skewed matrix data. *Journal of Multivariate Analysis, 167*, 181–194.

Melnykov, V., & Zhu, X. (2019). Studying crime trends in the USA over the years 2000–2012. *Advances in Data Analysis and Classification, 13*(1), 325–341.

Meng, X. L., & Rubin, D. B. (1993). Maximum likelihood estimation via the ECM algorithm: A general framework. *Biometrika, 80*(2), 267–278.

Misra, R. D. (1940). On the stability of crystal lattices. II. *Mathematical Proceedings of the Cambridge Philosophical Society,36*(2), 173–182 (1940). Cambridge University Press.

Punzo, A., & Bagnato, L. (2020). Allometric analysis using the multivariate shifted exponential normal distribution. *Biometrical Journal, 62*(6), 1525–1543.

Sarkar, S., Zhu, X., Melnykov, V., & Ingrassia, S. (2020). On parsimonious models for modeling matrix data. *Computational Statistics & Data Analysis, 142*, 106822.

Schwarz, G. (1978). Estimating the dimension of a model. *Annals of Statistics, 6*(2), 461–464.

Tomarchio, S. D., Gallaugher, M. P. B., Punzo, A., & McNicholas, P. D. (2022). Mixtures of matrix-variate contaminated normal distributions. *Journal of Computational and Graphical Statistics, 31*(2), 413–421.

Tomarchio, S. D., McNicholas, P. D., & Punzo, A. (2021). Matrix normal cluster-weighted models. *Journal of Classification, 38*(3), 556–575.

Tomarchio, S. D., Punzo, A., & Bagnato, L. (2020). Two new matrix-variate distributions with application in model-based clustering. *Computational Statistics & Data Analysis, 152*, 107050.

Viroli, C. (2011). Model based clustering for three-way data structures. *Bayesian Analysis, 6*(4), 573–602.

Author Index

L. Grilli et al. (eds.), *Statistical Models and Methods for Data Science*, Studies in Classification, Data Analysis, and Knowledge Organization,
https://doi.org/10.1007/978-3-031-30164-3

Printed in the United States
by Baker & Taylor Publisher Services